To Dad on his Birthday

Best Wishes

Peter & Chris.

CW01669321

THE ROYAL ENGINEERS

1101
26BT61
24

THE ROYAL ENGINEERS

T. J. GANDER

LONDON
IAN ALLAN LTD

Abbreviations

ADR	Airfield Damage Repair
AMF(L)	Allied Mobile Force (Land)
APB	Airportable Bridge
APC	Armoured Personnel Carrier
AVLB	Armoured Vehicle Launched Bridge
AVRE	Armoured Vehicle RE
BATUS	British Army Training Unit Suffield
BAOR	British Army of the Rhine
Bde	Brigade
BDT	Basic Driving Troop
CET	Combat Engineering Tractor
COC	Combat Operations Centre
CSB	Combat Support Boat
CRE	Commander Royal Engineers
EOD	Explosive Ordnance Disposal
Fd	Field (Squadron)
FFR	Fitted for Radio
FTX	Formation training exercise
FV	Field Vehicle
FWAM	Full Width Attack Mine
GV	Giant Viper
HESH	High Explosive Squash Head
HFB	Heavy Floating Bridge
HGOB	Heavy Girder Overbridge
HQ	Headquarters
HTT	Heavy Tracked Tractor
HWT	Heavy Wheeled Tractor
JSEODOC	Joint Services EOD Operations Centre
LCT	Light Crawler Tractor
LMD	Light Mobile Digger
LTT	Light Tracked Tractor
LWT	Light Wheeled Tractor
MC	Motor Cycle
MCAG	Mobile Civilian Artisan Group
MCE	Map and Chart Establishment
MCEG	Mobile Civilian Engineering Group
MCPG	Mobile Civilian Plant Group
MGB	Medium Girder Bridge
MHE	Mechanical Handling Equipment
MOS	Minimum Operating Strip
MTT	Medium Tracked Tractor
MWF	Military Works Force
MWT	Medium Wheeled Tractor
NI	Northern Ireland
ORATM	Off Route Anti-Tank Mine
PCS	Postal and Courier Service
RAC	Royal Armoured Corps
RAOC	Royal Army Ordnance Corps
RAOS	Repair of Airfield Operating Services
RCB	Regimental Commissioning Board
RE	Royal Engineers
Regt	Regiment
REME	Royal Electrical and Mechanical Engineers
RETDT	RE Training and Development Team
RHQ	Regimental Headquarters
RSME	Royal School of Military Engineering
Sec	Section
SDT	Specialist Driving Troop
SHAPE	Supreme Headquarters Allied Powers Europe
SMS	School of Military Survey
Sqn	Squadron
STRE	Specialist Team RE
TA	Territorial Army
Tp	Troop
V	Volunteer

ISBN 0 7110 1517 1

Published by Ian Allan Ltd, Shepperton, Surrey; and printed by Ian Allan Printing Ltd at their works at Coombelands in Runnymede, England

Preface

If I had know exactly what I was to become involved with when I was asked to write this book I would probably have declined. Now that the book is finished I am happy to have been asked, for it has certainly been an instructive and rewarding experience. The scope of the activities was such that at times I was wading through the mud of a West German field to catch up with an AVRE and at others I was seated in the quiet of what was to all intents and purposes a university devoted to military engineering. In between I have met many people and been shown many things, few of which I expected to see and even more that surprised me.

These reactions have been at the sheer diversity and range of tasks that the current Corps of Royal Engineers undertakes. The range of equipment and skills is very broad and although I have tried to place most of these within this book, I am sure that there are still some activities and items of kit that I have not included. To do full justice to the Sappers would require a book ten times this size but if there have been omissions or if certain sectors of the Corps feel that insufficient coverage has been given to them the fault is mine. It most certainly is not the fault of the many officers and Sappers of all ranks who have put up with my questions, demonstrated their skills and generally assisted me in every way they could. To all of them my thanks are due and I can only hope that this book goes some way to ensuring that the Royal Engineers get the due credit and recognition for the tasks they are carrying out in all parts of the world.

Terry Gander

(All photographs by the author unless otherwise credited).

Contents

Introduction

The Royal Engineers always have been special. Even in an army as well endowed with talent and powerful personalities as the British Army the Engineers have stood out as being more talented and personable than most. From the ranks of the Royal Engineers have come some remarkable men who have made their mark on history in many ways, for the soldiers who enter the ranks of the Sappers usually do so with a sense of vocation or at least follow a sense that they are embarking on a career that will best suit their various temperaments. In the main they have been well rewarded with a military life tht has sent them to all corners of the globe and presented them with a seemingly never-ending series of challenges, opportunities and consequent rewards. From the ranks of the Royal Engineers have come a disproportionately high number of high-level commanders and men who have influenced civilian and military life and matters. Their impact has been more far-reaching than many would realise. Many high-ranking commanders in both world wars were Sappers and many major civil engineering projects have been planned or carried out by men who learned their skills in the Royal Engineers. A small measure of their impact can be determined from the fact that it was Royal

Below:
Mobile cranes lifting a mine flail onto a Churchill tank chassis, one of several mine-clearing methods employed by the Royal Engineers during World War 2, but also used on occasion by other arms. *MoD*

Engineers who irrigated India, played a major part in the foundation of the Ordnance Survey, entombed Napoleon on St Helena, planned the drainage of London, gave their names to mountains such as Mount Everest, burned the White House in Washington, discovered the sources of the Nile, designed the Albert Hall and manned the world's first underwater guided weapon. If all that were not enough their influence on the development of the modern British Army has been profound. The Royal Engineers acted as the foundling father of many modern corps and regiments. From the Royal Engineers have sprung the Royal Flying Corps (and from the RFC the Royal Air Force and Fleet air Arm), the Royal Signals, the Royal Corps of Transport and the Royal Electrical and Mechanical Engineers, the latter two have taken over many of the functions that were once the perogative of the Royal Engineers. They taught the Royal Navy how to handle submarine mines and handed over to the present Property Services Agency many of their one-time responsibilities. And they have had their hand in many other aspects of military and civilian life.

The present Corps of Royal Engineers can officially place the date of its present establishment back to 1716 when King George I issued a Royal Warrant to establish a Corps of Engineers as part of the Board of Ordnance. In fact engineers had been a part of the British military establishment for centuries prior to that date. When William the Conqueror landed in 1066 he brought with him his own King's Engineer, but the first 'British' military engineer is generally acknowledged to be the Bishop Gundulf who was responsible for a number of feats of construction such as the White Tower of the present Tower of London and a tower at Rochester Cathedral. The latter forged a link between Rochester and later generations of engineers that remains to this day. During the Middle Ages the duties of the King's Engineer were carried out by a succession of noblemen who built bridges, castles and towers throughout the realm – with a growing tendency toward coastal defences, such as those built during the reign of King Henry VIII.

By the end of the Middle Ages the art of the military engineer had become well established as the invention of gunpowder reduced warfare to a series of sieges of well-protected fortresses. Indeed, the siege and the subsequent operations attendant upon it led to the engineers' nickname of 'Sapper'. This name was derived from the approach trenches that were dug towards the walls of a besieged fortress, a trench that was dug in a zig-zag form known as a 'sap'. Thus the digger of the sap became a sapper and the name has stuck.

The expansion of British interests in the 17th century meant that soldiers were spread around the world and with them went the engineers. Specialist military companies of engineers were established in India and others were involved in wars in Canada. At this time the engineers, such as they were, came under the control of the Board of Ordnance which concerned itself not only with the engineers but also with the artillery. Thus when the Royal Warrant was issued in 1716 it established not only the Corps of Engineers but the Royal Regiment of Artillery and to this day the Engineers and the Gunners share the same regimental colours (dark blue and red).

The initial establishment of the new Corps was modest – only 28 officers. For their labour force they had to rely upon civilians but with these they built a chain of new barracks and constructed many civic buildings including those that still grace the centre of Dublin. But the 18th century was far from peaceful and engineers carried out their military functions in all parts of Europe and elsewhere. In 1741 a significant step forward was made by the establishment of the Royal Military Academy at Woolwich, set up specifically to impart a military education for new officers joining the Engineers and Gunners.

It was the Great Siege of Gibraltar – from 1779 to 1782 – that provided the Corps of Engineers with its first European work force subject to military discipline. This was the Soldier Artificer Company that dated from 1772 and was formed in Gibraltar specifically to improve and construct the defences that were to withstand the Great Siege itself. The efforts of the Corps of Engineers and the Soldier Artificer Company during the siege were such that in 1797 the Corps was granted the title of Royal and thus became the Corps of Royal Engineers.

The new Corps was still without a work force directly under its own military control. The original Soldier Artificer Company was expanded to two companies and in 1787 they were joined by a number of new companies (six in all) based in the United Kingdom and known as the Royal Military Artificers. These new companies were commanded by Royal Engineers officers and by 1797 the original Soldier Artificer Companies had been absorbed by the new Royal Military Artificers. More of the new companies were formed for overseas service until by the end of the Napoleonic wars the Royal Engineers and Royal Military Artificers were serving all around the world.

During the Napoleonic Wars the Royal Engineers made their greatest impact during the Peninsula campaigns. Their fortunes varied from the frantic assaults on Badajoz to the tremendous efforts produced by the construction of the Lines of Torres Vedras that were built to protect Lisbon. The construction of the Lines was one of the major feats of military engineering of the era and was carried out almost entirely by civilian labour under the guidance of the Royal Engineers and the Royal Military Artificers.

After 1815 the number of Sappers in the British Army fell away as the nation entered a period of relative peace but by then the Royal Military Artificers had become the Royal Sappers and Miners. The change was made in 1812, partly as the result of recommendations made by the Duke of Wellington following his experiences of military engineering strengths and weaknesses during the long Peninsula campaigns. There had been times when the Royal Engineers and the Royal Military Artificers had shown themselves to be lacking in the necessary skills and equipment involved in many aspects of the campaign and the establishment of the Royal Sappers and Miners was instrumental in the formation of a skilled body of specialists who could train and carry out all the many forms of military engineering from the assault to everyday duties. This move allowed the Engineers to become far more professional and better equipped than had hitherto been possible. The Royal Sappers and Miners remained in being under the leadership of the Corps of Royal Engineers for extended periods to allow the formation of proper drills and procedures to become established. New equipment was designed and produced to be ready for use at all times rather than the hasty field improvisations that had been the norm until them. The officers of the Corps of Royal Engineers were able to spend more time investigating the finer points and the sciences of their profession to the extent that their services were frequently called upon for civil projects. The Corps of Royal Engineers had been involved in the affairs of the Ordnance Survey for many years before the accurate mapping of England was commenced in 1787 and long before the Survey Companies of the Sappers and Miners were formed in 1824. The Corps of Royal Engineers and the Royal Sappers and Miners were heavily involved in the great programme of fortification construction that extended from Corfu to the coasts of England and Ireland during the mid-1800s (and extended until the end of the century), while abroad Engineer officers overlooked the construction of canals – the Rideau Canal between Montreal and Kingson in Canada was one of the major construction projects of the 1820s.

Throughout the first half of the 19th century the division of the Corps of Royal Engineers and the Royal Sappers and Miners was maintained until the old Board of Ordnance was abolished in 1855. The following year the two separate bodies were united under the title of the Corps of Royal Engineers and at long last the officers of the Corps had direct command of the men whom they usually had to call upon for the implementation of their various engineering projects.

What history now regards as the high points of the Victorian era were important times for the future of the Royal Engineers. In the long term perhaps the greatest influence was felt by the integration into the Corps of the engineer elements of the old East India Company during the period that followed the upheavals of the Indian MUtiny in 1857. Up to the Mutiny the East Indian Company had virtually run its own private army complete with all the trappings of a state body including a Staff College, and an engineer arm every bit as proficient in its duties as the Royal Engineers. There is no space in this book to make a full account of their accomplishments over the years but suffice it to say that it was companies of the Bengal Sappers and Miners who built the massive Great Trunk Road across India from Calcutta to Peshawar. That was not the only road they built, for much of India's road infrastructure can be traced back to those able Victorians. They also ran a silver mint, dug canals, carried out a survey of the nation that is still used today, built docks and warehouses and – apart from the mundane building projects such as barracks – it was an Indian engineer who constructed the mighty edifice of St Pauls Cathedral in Calcutta.

Signs of the future could be seen in the fact that Indian engineers built India's first railway line. Between Bombay and nearby Kalyan snaked the first steel rails that were later to unite a sub-continent and the culmination of these early efforts was later to be seen in the building of the line from the Indus to Quetta, one of the major railway building feats of the 19th century.

The officers of the Indian engineers merged with the Royal Engineers in 1862 and the stage was set for a major expansion of the Corps' work and influence. This expansion was brought about by the nature of the colonial campaigns of the period and also by the influence of the men involved – characters by any definition who were to leave their mark on the future of the Corps. At their head came 'Chinese' Gordon but just as important were Bindon Blood, Robert Napier, 'Buster' Brown (who built the railway to Quetta), Colin Scott-Moncrieff (who irrigated Egypt and laid the foundations of that country's modern water systems), and Alexander Taylor who was one of the old Indian engineers. Perhaps the most important of all the engineers of this era was Kitchener who started life as a sapper and who later became the best known soldier of his generation and whose face dominated the recruiting placards of the Great War years.

It was these men and their contemporaries who guided and led the Corps at a time when the world

Above:
A Free French Sherman tank making use of a Churchill Ark Mk 1 to cross a sea wall. The Ark was a special piece of combat engineer equipment developed for use by the 79th Armoured Division and used by the Royal Engineers and some other arms. *MoD*

saw the full force of Victorian technology. Throughout this period whenever some innovation was introduced to the military scene the Royal Engineers were usually involved. Typical of these innovations was the growth in communications, usually involving telegraphy of various kinds. In 1870 the first Telegraph Companies were formed within the Corps and grew in numbers and influence to the extent that by 1885 the Nile Expedition had on hand a whole battalion.

In 1867 the first of many Railway Construction Companies were in action in Abyssinia. About the same time a new science was introduced into warfare in the form of the underwater mine for harbour defence. These minefields were carefully laid in likely locations and detonated under remote control from shore stations. These minefield shore stations came under the control of the Corps' new Submarine Mining Companies from 1871. The Royal Engineers looked after the shore installations and also manned the defence electric lights, the searchlights that covered the approaches to harbours and defended locations. One item of equipment that the Corps used during this period was the Brennan Torpedo which can be regarded as the world's first guided weapon – it was guided towards its target via wire-carried commands. The Corps retained these duties until 1905 when the mining aspects were handed over to the Royal Navy leaving only the defence electric lights in the hands of some Royal Engineer Fortress Companies and some militia units around the United Kingdom.

The Corps was involved with all the various bush wars and colonial campaigns in which the

British Army became involved throughout the late Victorian period. There is again no space to mention them all, but one of the most famous was the saga of Rorke's Drift during the Zulu war. It should be remembered that the leader of the defenders of this little outpost was one Lt Chard, a Sapper officer commanding a ferry at that particular location when it all happened. His richly deserved VC was but one of many won by Sappers.

Ballooning was another innovation that drew the attention of the Corps during the late Victorian period. The Corps had opened a Balloon School at Chatham in 1888 and by 1890 a Balloon Section was in existence. During 1892 a factory to produce balloons was established on Laffans Plain and named the School of Ballooning. In 1905, following on from an active period of service in South Africa during the Second Boer War, the school was moved to a new location near Farnborough where a few years later it formed part of the initial framework of the organisation that now bears the title of the Royal Aircraft Establishment. The Balloon School remained a Royal Engineer unit until 1911 when it became the Air Battalion RE, only to change its title once again a year later when it ceased to be part of the Corps and became the Royal Flying Corps (Naval and Military Wing). This unit went to war in 1914 and by 1918 the Royal Air Force and Fleet Air Arm were formed from the greatly expanded aerial fleets that had grown from the intial tiny battalion. The Royal Air Force's links with the Royal Engineers have yet to be broken for No 1 Squadron, Royal Air Force still likes to trace its history back to the Balloon School days. Today the squadron flies Harriers, an aircraft that obtains much of its ground support from the Royal Engineers in BAOR.

The balloons in South Africa were but one manifestation of the expanding numbers and scope of the Corps. Another example of this can be seen in the formation of a steam road traction unit for service during the Boer War. This was 45 Company RE but soon after the arrival of the petrol engine involved the Corps in yet more road traction duties to the point where some of the Corps' responsibilities were handed over to the new Army Service Corps, much later the Royal Corps of Transport. But while some duties involving road transport were hived off others, including maintenance, were retained until the formation of the Royal Electrical and Mechanical Engineers (REME) in 1942. Thus the Corps had its part to play in the formation of two more British Army corps.

The Royal Engineers in 1899

The Regular Army

The Field Troop – a mounted detachment
Bridging Battalion – two Pontoon Troops
Eight Field Companies
The Field Depot – two Field Parks and a Training Depot
The Balloon Section and Depot
The Telegraph Battalion – one Division at Aldershot, one Division with the London Postal Telegraph
Eighteen Fortress Companies plus a West Indian Fortress Company
Eight Depot Companies
Twelve Submarine Mining Companies and a Coast Battalion of eleven sections (Defence Electric Lights)
Five overseas Submarine Mining Companies – Ceylon, Hong Kong, Mauritius, Singapore and the West Indies

Two Railway Companies
Four Survey Companies

The Auxiliary Forces

Two Militia Battalions – the Royal Anglesey and the Royal Monmouthshire
Twelve Militia Submarine Mining Divisions
The Engineer and Railway Volunteer Staff Corps
Twenty Volunteer Battalions
Seven Volunteer Submarine Mining Divisions
Corps of Electrical Engineers (DEL)

Left:
A Sherman Crab Mk 1 mine-clearing flail in action. *MoD*

The full scope of the Royal Engineers in 1899 can be seen in the accompanying table and an indication of its duties and types of unit will be provided by its content and variety.

One major activity of the Corps that was later to be handed over to an entirely new body was the telegraphy function. This activity had embraced all aspects of field communications of the period from flag-waving to the use of rudimentary telephones and the telegraph. Sappers went into action in South Africa with all the specialised paraphernalia of wire-laying waggons and the like which were still in use when the BEF crossed to France in 1914. The years 1914-1918 saw a dramatic increase not only in the scale of Army communications but also in the technology. The Royal Engineers had been involved in ground-to-air radio links as early as 1911 and were involved in many of the later improvements in equipment and operating methods. In fact the art of military communications reached the stage where it was decided to form a corps to specialise in the subject and the Royal Engineers handed over their communication responsibilities to the new Royal Corps of Signals in 1920.

One further aspect of the Corps' activities during the run-up period to the Great War was the unsung efforts of the various 'stores' and 'works' companies. The organisation of these was based on the experience gained from the supply and maintenance of the various Submarine Mining Companies tht required their own special services. Over the years these service companies grew gradually but during the Great War they expanded out of all recognition.

By 1918 there were 1,832 Royal Engineer units of all kinds manned by a complement of over 330,000. All this had all been built on the relatively small Corps structure that went to war in 1914. The expansion of the Corps during the Great War was largely unforeseen and was carried out on an ad hoc basis in which the versatility and flexibility of the Corps was frequently stretched to the full.

The Survey Companies were among the first to expand in number and were soon involved in the new science of producing military maps using the novelty of aircraft photographs. This was not the only new science of the period for the Corps was soon involved in the horrors of chemical warfare, both in defence and attack. It was a Sapper who gave his name to the Livens projector used to fire canisters of chemical substances at the enemy's trenches; another name that was to pass into the English language came from the sapper who gave his name to the Nissen hut.

The demands of the Western Front soon produced new types of Royal Engineer unit. There were Tunnelling Companies to drive mines and galleries under the trench systems of the front, Forestry Companies to fell timber and the new Field Survey Battalions that fed the army with new maps almost every day. The Field Companies supplied the army with most of its requirements, including the novel weapons that trench warfare demanded and things like pumps which made trench life a little more tolerable. Where the required equipment or services were not available the Field Companies produced them. Hand grenade factories were established in rear areas and building were converted into bath-houses for troops returning from the front. Tracks and tramways were built to carry men and supplies, and the materials that passed through their hands were soon being handled in numbers where millions were commonplace.

As has already been mentioned the Royal Engineers were involved in all aspects of chemical warfare from the start and the Special Brigade RE not only manufactured war gases but also produced the means to deliver them to the enemy. The Livens projector was one; others included gas cylinders and the first attempts at gas shells. The flamethrower was another Sapper innovation following its introduction by the Germans and the Corps was also involved in the development of the first gas masks.

Chemical warfare was not the only technical innovation of the Great War, albeit an important one. Even more important was the introduction of the first tanks in 1916. The Corps was not directly involved in their development and introduction but many of the men involved were Sappers. A Lt-Col Swinton had put forward the idea of the tank as early as 1914 but he had to wait until 1916 before his foresight was rewarded. Swinton then commanded the Heavy Section, Machine Gun Corps – the first tank unit – and when this became the Tank Corps in May 1917 it was commanded by another Sapper, Brigadier Elles. Other engineers involved with the tank at this period included Major Inglis, who designed some of the first tank-carried bridging, and Major Martel, who was later to produce some of the small vehicles known as 'tankettes' that were to lead to the tracked carriers of 1939-45.

Away from the Western Front with its ever-expanding array of Artisan Works Companies, Water Boring Sections, Land Drainage Companies, Road Construction and Quarrying Companies (there was even a War Dogs School RE) the railway units multiplied. By 1917 the original five railway units had grown to 130 and their duties involved not only running railways but attending to the construction, operation and repair of harbours, docks and bridges to say nothing of the railway stock. The Corps also undertook the smooth running of men and material along the railways with the worldwide Transportation and Movement Control Service. Thus Sappers were to be found smoothing the passage of men and supplies not only in France but in Salonica, Mesopotamia, all through the Middle East and, of course, India. This quiet function was carried out by the Royal Engineers until it was handed over to the Royal Corps of Transport in 1965.

The railway units operated all over the place sometimes running services that would have been difficult in times of peace. The Corps found itself running railways in Italy, Greece and eventually Russia. There was even a Sapper navy of sorts operating in the Persian Gulf operating whatever vessels were available, including some pleasure boats that had once worked on the River Thames and were sailed out to their new station by Sapper crews.

Once the Great War was over the Corps and its myriad companies and units once more shrank to something resembling its prewar size. As the numbers dwindled so did its range of duties: some responsibilities were handed over to other corps, as has been related, but life in the Royal Engineers was seldom dull during the interwar years. At various stations around the Commonwealth and Empire Sappers still found much to do, and even at home the reorganisation of the interwar years still gave rise to the need for new barracks and other installations.

By the late 1930s the whiff of expansion was once more in the air. One area was a left-over from the Great War when the home defence of the United Kingdom had been assisted by the formation of searchlight units as an outgrowth of the old DEL duties. But the new searchlight units pointed their beams skywards instead of out to sea and by the late 1930s these duties were once more assumed as the number of anti-aircraft searchlight units grew. In fact they grew to such an extent that they were too numerous for the Corps and this function – both anti-aircraft and DEL searchlights – was gradually handed over the Royal Artillery.

Other plans for the future were well under way by the late 1930s. New combat engineer equipment, from pontoons to bridges, was in the

Right:

One postwar special engineer vehicle that didn't make it, the FV437 Pathfinder, a member of the FV432 family. This was an amphibious engineer vehicle fitted with a small dozer blade and other special equipment, including the anti-personnel mine projector seen over the hull. Only a prototype was made and the project got no further.

pipeline. By the time World War 2 started, the Corps was mechanised and for the first time when they moved into France they would do so without the horse.

The scope of RE activity during World War 2 is too large to be mentioned here – just to list them would fill this book. Needless to say new duties were found for the Corps, one of the first starting during the Battle of Britain in 1940 when the disposal of unexploded bombs became such a priority that new Sapper units were formed to deal with them. This function has been with the Corps ever since. Another new duty was that of Armoured Assault Engineer in which a whole new family of specialised armoured vehicles moved forward with armoured units to clear the way and 'take out' strongpoints and defended positions. Land mine warfare grew out of all recognition from the early days when mines were simply popped into ready-made holes in the ground. In the place of such early methods grew massive minefields covering huge tracts of terrain – and with the growth in mine-laying came the growth of mine-clearing using mechanical devicers such as ploughs, flails and rollers some of which were also used by other arms.

New equipment had to be mastered and used in action. Here again a list would be too long for completeness but items that must be mentioned include the Bailey Bridge, a whole new array of demolition and other devices, and items such as new assault boats and pontoons for river-crossings.

Sappers served worldwide, joined in many theatres – as they were during the Great War – by their counterparts from the Commonwealth and Empire. They fought and worked wherever they were needed and that meant everywhere. The old cry of 'Call for the Sappers' was increasingly heard in all corners of the Globe between the years 1939 and 1945 but not all the Sapper efforts were in the front line. Behind the lines and in such places as Gibraltar, Malta, East Africa and Egypt the Sappers carried on their job of helping the army to live by digging tunnels, wells and watercourses. They supplied water and electricity, ran the railways and organised and operated the docks. Their assault bridges often had to remain in place for months until proper bridging could be supplied, and they also built airfields and landing strips for the Royal Air Force.

During World War 2 there are several Sapper achievements that stand out as worthy of special mention, some of which now hold honoured places in Corps history. The clearing of the minefields during the Second Battle of El Alamein stands out as one particular highlight but it was surpassed by their involvement in the Normandy landings of June 1944. The Corps not only cleared the way ashore for the assault craft but once ashore their new armoured vehicles cleared the way for the rest of the forces to move inland. The

Sappers were also operating from the sky as well as the sea. The first combat engineer airborne units went into action during the 1943 Sicily operation and they too were in action in Normandy in June 1944.

Once ashore the Sappers built new airstrips, organised the supply of fuel and kept the small roads of Normandy open and able to bear the huge weight of the traffic that poured through them in the days after the landings.

If all this was not enough the Rhine Crossings lay ahead. Again, this was very much a Sapper operation. In the end the crossing was made successfully but only by the dint of preparation and planning before the event. Assault bridging, boats and all manner of supplies and equipment had to be moved to the exact location where they were required and the amounts involved were prodigious. Compared with all this work the actual building and emplacing of the bridges was over in a trice but once the Rhine was crossed the Army was free to move across Germany and the end came in May 1945.

In the Far East the Royal Engineers worked with the Indian Engineers to fight the last campaign of the old Indian Army in Burma. They covered the withdrawal from Burma and then retraced their steps all the way back to Rangoon. During the Burma campaign the Sappers frequently had to work under dreadful conditions but they continued to build roads, bridges and airstrips, supply water and operated supply airstrips. Burma was not a campaign where the internal combustion engine could play a large part apart from in the air. Frequently the mule and the buffalo were all that could be used and at times Sappers even organised their own river supply routes to bring forward supplies and equipment. All the time the tropical climate and environment was at its worst but the Sappers endured it all until the end.

The postwar years were far from peaceful. Malaya and Korea still lay ahead in 1945 but the initial task was to clear up the mess and return to normal. Sappers also returned to their homes in large numbers and by 1946 the huge numbers of the war years had once more returned to a stage where peacetime levels could at least be discerned. But the numbers were never allowed to dwindle to prewar levels. The threat of nuclear war kept a steady stream of conscripts churning through the Royal Engineers and many of them were provided with more than enough work to do during their two-year National Service.

While the Korean War erupted at the end of the 1940s, it was the campaign in Malaya that gave a more meaningful indication of the nature of postwar soldiering. The years after 1945 saw a gradual withdrawal from Empire with all its attendant upheavals, civil unrest and internal security problems. Throughout it all the Sappers have been deeply involved and nowhere more than when the civil unrest reached the United Kingdom with the troubles of Northern Ireland. There the Sappers have built protected barracks and taken their part in the constant patrolling. It has not been an easy task for the Sappers in Ulster but the task was no easier in Aden, Cyprus, Palestine or any of the other postwar trouble-spots.

Throughout all these years the Royal Engineers had to stand at the ready in Germany as part of the British Army of the Rhine. There they remain today and one of the objectives of this book is to relate how they carry out those duties. But one thing that must be borne in mind while reading of these military duties. There are several aspects of the Royal Engineers that seldom reach the headlines but are nonetheless important functions of a Corps that quietly carries the flag to all parts of the world.

The Royal Engineers do not always carry out their myriad functions with a purely military objective in mind. They also perform many services to civilian populations and communities, both in time of trouble and peace. Throughout the world Sappers have always built roads and bridges, established water and electricity supplies, carried out surveys and produced maps, and constructed public utilities and buildings of all kinds. Much of this work has gone unnoticed by the world at large but to those who look for their involvement it will be noted that whenever there has been some form of natural disaster such as an earthquake or famine the Royal Engineers are often there to build, supply and generally assist relief operations. For the people affected the Sappers are always welcome and the services that the Corps can provide in such circumstances have often made the difference between life or death. Disasters are not the only source of Royal Engineer assistance however, for their services are frequently requested by many nations, usually those that were once part of the old Commonwealth and Empire, keeping alive links that once appeared to be broken.

Finally so much mention has been made of work that it would be easy to forget that Sappers can play as hard as they work. Right from the old days of the Raj Sappers have always known how to relax and this talent survives to this day. Be it sailing, adventuring or just plain enjoyable sport the Royal Engineers have always been involved to the full. Today their sporting record within the Army is second to none and a brief look through the Olympic lists over the last 50 years will soon show how the Corps has maintained this fine sporting tradition. It should not be forgotten that the Corps won the FA Cup in 1875!

1 Roles and Organisation

The main role of the Corps of Royal Engineers is to provide combat engineer support to the brigades and divisions of the army. In broad terms this support is provided by 15 Field Engineer regiments, an Amphibious Engineer regiment, an Armoured Engineer regiment, a Commando squadron and a Parachute squadron. These combat units make up some 70% of the manpower strength of the Corps which numbers about 13,800 men of all ranks.

This quick outline is only part of the story, for to support these units there are many other units, most of them specialist, and to add to them are the numerous Territorial Army units that wear the Corps cap-badge. The specialist units include the Explosive Ordnance Disposal (EOD) Regiment, the Engineer Support Group (the Military Works Force), the Survey Branch and the Postal and Courier Service. There are numerous other units scattered all over the world including the Mobile Civilian Engineer groups based in Germany, a Specialist Team Royal Engineers (STRE) in Washington, the Queen's Gurkha Engineers and all the various training establishments.

All of these will be covered in the main sections of this book, but the Corps' function of providing combat engineer support to the Army requires a closer definition. It has been succinctly stated that combat engineer support means helping the Army to live, work and fight while at the same time denying the same facilities to an enemy. It would be difficult to summarise the combat engineer support role better but a listing of the services would perhaps emphasise the point.

The list includes: road and route construction and maintenance, mine-laying and clearing, the

Below:
A FV432 storming ashore after a ferry trip on a Class 16 airportable bridge raft.

construction of field fortifications of all kinds, bridge building, water supply, demolitions, the construction of landing strips and helicopter landing pads, and so on. Many more items could be added but the main point to emphasise is that the Royal Engineers is a very versatile Corps which can turn its hand to virtually anything that the army might require, and they have done so on many occasions in the past.

The very term engineer is derived from the latin *ingeniarius,* which was once defined as 'a person skilled in the art of constructing defences, more gifted than others, and perhaps a genius'. No doubt some modern combat engineers would like to use that definition for themselves but perhaps a more accurate definition for today would be that the combat engineer is today a soldier with three abilities and these are all reflected in the manner that combat engineer support is provided.

First and foremost the combat engineer is a fighting soldier, fully conversant in all the various infantry and small-arms skills required for him to act as a fighting unit on the battlefield. Secondly he is a combat engineer fully able to handle all the various items of combat engineer equipment. Last, but not least, he is a tradesman with a specialised artisan skill that can be turned to provide virtually any trade support that might be required.

To utilise these complex skills the Corps is organised into regiments each consisting of a

Left:
FV180 CET prepared for swimming and with recovery anchor in place.

Below:
AVRE 165 with dozer blade raised.

number of squadrons, each regiment being roughly (very roughly) equivalent to an infantry battalion in size and form. However, having said that, there are significant differences between units in Germany and those in the United Kingdom. The army in Germany, or the British Army of the Rhine (BAOR), is at the sharp end – it trains and stands ready as part of a much larger defence force organised under NATO. The units in BAOR are combat units in every sense of the word and are ready to move off into the field to fight at very short notice. The same may be said of some of the units based in the United Kingdom and elsewhere but they have to move to Germany before they can take up their combat status and are thus organised somewhat differently to the BAOR units.

There is one further item to mention at this point and that is that some Royal Engineer units have two manning and organisation levels for war and peace. This book will deal mainly with the peace time status unless otherwise mentioned, for with many units their war status will depend on a number of factors such as actual location at the outbreak of hostilities, the anticipated degree of combat, manning levels at any one particular time, and so on.

To give an overview of the Corps in BAOR, each

of the three armoured divisions in 1(BR) Corps has one Field Engineer regiment. This tidy arrangement is backed up by a further two Field Engineer regiments which are under Corps control and can be seen as equipment-holding regiments to support rear areas. Also under 1(BR) Corps control are the two specialised combat engineer regiments – the Amphibious Engineer Regiment and the Armoured Engineer Regiment.

For every combat engineer formation within the Corps there is usually a support unit of some description. All the 1(BR) Corps regiments are supported by a special support squadron and this in its turn can call upon the support of a support group consisting of base workshops and engineer equipment parks. Other specialised Corps units exist within BAOR including a lighting troop, a field squadron that deals almost exclusively with field support for the Royal Air Force's Harrier squadrons, two postal and courier regiments and a map-making squadron. There are also the Mobile Civilian Engineer Groups which have a peacetime function of maintaining the road and building infrastructure of all kinds of facility within BAOR. There is also the Combat Engineer Training Centre.

Inside Germany but outside the normal remit of BAOR is the Berlin Brigade where a single two-troop field squadron provides local support. Berlin is also unusual in that it is the home of one of the few Territorial Army units outside the United Kingdom, a single field troop (V).

In the United Kingdom there are four Field Engineer regiments but here there are no tidy allotments of regiments to divisions. Each of the United Kingdom field regiments does have its part to play in time of war but there is no tidy way to describe these allocations. Regiments and squadrons each have their wartime allocation with some moving the BAOR, some to the NATO flanks to support the Allied Mobile Force (Land) – AMF(L) – some for airfield damage repair in Germany, some for home defence, and so on.

One of the more specialised aspects of the Royal Engineers is the Explosive Ordnance Disposal Regiment based in the United Kingdom. As will be related later this regiment is currently expanding and in addition to its regular squadrons also has a number of volunteer Territorial Army squadrons. The Military Works Force also has its base in the United Kingdom although it only rarely comes together as one unit and with this force are the many less well-known specialist volunteer units that come under the Corps banner. The volunteers also provide an entire brigade for airfield damage repair but they make up less than six Field Engineer regiments (V) and an Engineer Brigade (V). The exact roles of these regiments again cannot be precisely outlined as they vary so much.

Both survey and postal and courier regiments are based in the United Kingdom but it must be stressed that the United Kingdom acts as the main training and support base for the various Corps units in BAOR. The bulk of the reinforcements in an emergency would come from the United Kingdom and in time of peace the home base acts as the main training location for all the various schools, training establishments and depots that the Corps requires.

Outside the United Kingdom the main centre of Corps activity is still the Falkland Islands although the current level of activity is now running down compared to the frantic efforts of the campaign and post-campaign involvements of 1982 and immediately after. Various units still visit the islands on a roulement basis to provide support for the garrison, as they do for Belize, and a sizeable body of the Corps' manpower is stationed there – there will be an EOD involvement for years to come.

The Queen's Gurkha Engineers provide engineer support for the Brigade of Gurkhas away in Hong Kong and Brunei and there is also a Gurkha Independent Field Squadron based in the United Kingdom for the support of a home defence brigade.

There is also a Royal Engineer presence in Cyprus and another in Gibraltar where the only reminder of the old Fortress Companies (RE) still survives. In Washington DC there is a highly-specialised unit dealing with co-operation between the various American agencies concerned with space satellites and their use for survey and other purposes.

This is not an exhaustive listing of all the various aspects of the modern Corps of Royal Engineers but it does provide an overall view of the problems that face an author who has to outline all the many and various activities, organisations and equipment types that go with a description of the Corps today. After some thought it has been decided to outline the various organisations and equipments by dealing with them by functions in the following chapters:

2 Combat Engineering

The term combat engineering covers a wide range of military activities – not just digging holes. While this time-honoured activity remains, it is but a tiny part of the Corps' combat engineering activities.

The modern army is a mobile army and means that combat engineering has to be carried out rapidly and with what is to hand. It may be possible in a few circumstances to pre-position combat engineering equipment and materials but in any future war these opportunities will be few. Men, equipment and materials have to be taken to a location, used quickly and moved on as soon as possible. Consequently the modern combat engineer is a highly mobile being who is taken to his working site by a machine; once on site he usually uses some form of machine to aid his task and the task completed he moves on using some form of automotive transport.

The combat engineer will also be operating in a hostile environment where attack might be expected at any time and even when on the move the attack threat does not diminish. Some form of protection has to be provided therefore and the only form of protection of any use on the battlefield is armour. Thus the Field Engineer regiments operating in Germany use tracked and armoured vehicles, usually FV432s or FV103 Spartans. Not all the engineer equipment can be as comprehensively armoured as these two vehicles but the combat engineer can usually

Below:
'Turretless' Ferret scout car used by 39 Field Engineer Regiment.

expect to travel into action in a FV432 and vehicles such as the FV180 Combat Engineering Tractor (CET) do provide some measure of protection for their crews. However much of the earthmoving plant is unarmoured. A field regiment operating in BAOR has on its books about 100 'A' vehicles, nearly 150 'B' vehicles and 40 'C' vehicles to say nothing of the various trailers and items of ancillary equipment that go with them.

BAOR Field Engineer Regiment

The types of task carried out by a Field Engineer regiment operating in Germany are many. They include the rapid erection of field fortifications and obstacles such as anti-tank ditches (still a very formidable obstacle to any armoured fighting vehicle), the provision and construction of assault bridging, the digging of defensive positions such as weapon or tank slots and trenches, route construction or improvement, clearing obstacles and mine clearance. Combat engineering also means denying mobility to an enemy so also involved are mine-laying, demolitions and the erection of barbed wire and other obstacles.

To carry out these and the many other battlefield duties is the task of the Field Engineer regiments, be they based in BAOR or the United Kingdom. The major difference between the BAOR regiments and those in the United Kingdom is that the home-based regiments use wheeled transport instead of tracked and sometimes their establishments are smaller. But of course not all BAOR equipment is tracked – while the BAOR combat engineers usually travel in tracked vehicles their logistic and other support functions are wheeled as well, so the differences between BAOR and UK-based engineers equipment is not that great.

Above:
Washing down a FV432 after a training exercise.

Starting at the top, each of the Field Engineer regiments has a regimental headquarters (RHQ), and three field squadrons plus a field support squadron in support. There is also a REME workshop in support.

Each of the field squadrons has its own squadron headquarters commanding three field troops and a support troop. Taking the field troop, which is the basic engineer sub-unit, down to its final level there are four field sections under a troop headquarters. (United Kingdom-based field squadrons have three field sections in a troop).

Unfortunately there is no neat method of describing how all these engineer units and sub-units will tie in with the armoured division organisation. It would be too neat to say that each field squadron would associate with one of the three brigades within each division, for battlefield organisation does not work like that. The units within 1(BR) Corps would, in time of war, remove themselves from their usual administrative trappings and assume a new organisation based on the battle group and combat teams. There is no hard and fast establishment for a battle group for they are formed to suit the particular combat task that a battle group commander has to assume. Thus although the battle group is supposed to be a balanced mix of infantry, artillery and engineers the strengths of each of these three will vary from group to group. Some battle groups might require the support of two field squadrons while another will require but a single field troop.

Thus to describe the organisation of the Field Engineer regiment it would be as well to start with the basic engineer sub-unit, the field troop. Each BAOR field troop has four field sections, each made up from a single vehicle, a FV432 commanded by a corporal and carrying a further eight or so men, including the driver of the FV432. If required each field section can be detached for small tasks such as demolition for short periods but most of the time it will remain under the control of the troop headquarters.

The troop headquarters is usually commanded by a junior officer with a staff sergeant, a reconnaissance NCO and a small logistic element. The troop headquarters is based on the use of three FV103 Spartans that double as control and communication vehicles and reconnaissance vehicles. Each troop has at least one Bar Minelayer.

Each field squadron – see accompanying diagram – has its own support troop to supply earthmoving and other equipment and support for the field troops. The equipment of the support troop is under the control of the field squadron headquarters and is assigned to the field troops as required. The main strength of the support troop consists of five CETs which are assigned complete with their crews to wherever they are needed. Other equipment that would be assigned as required consists of a single light mobile digger (LMD) with its rapid trench-digging capabilities and a light wheeled tractor (LWT) with its hundred-and-one uses. The support troop also has a significant refuelling capability with a 4-tonne truck equipped with fuel tanks and refuelling equipment. The support troop Stalwart acts as a general load-carrier and the Ferret scout car is used for reconnaissance.

Going further up the line the field squadron headquarters has two FV105 Sultans for the officer commanding and his immediate staff along with Land Rovers and the seemingly inevitable Ferret scout car. Note that there are three motorcycles held at the headquarters and there are no less than six 4-tonne trucks for general load-carrying and support purposes.

Organisation and Equipment of a Field Squadron in BAOR

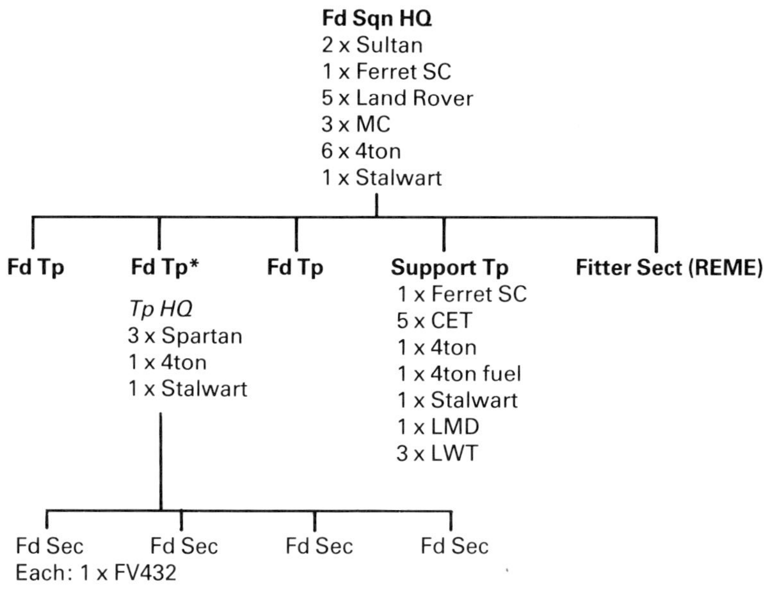

*Note: UK field troops have only three field sections.

One section that is very easy to forget when dealing with the field squadron headquarters and indeed every level and type of Royal Engineer organisation is the fitter section, manned by the Royal Electrical and Mechanical Engineers. Using their FV434 repair vehicles these REME personnel keep the whole squadron running by front-line repairs and even such items as engine packs for the FV432s can be changed in the field. Without the REME the Royal Engineers would be unable to remain in the field for very long.

The field squadron commander would probably not be with the squadron during operations as he would be with his tactical headquarters consisting of a single FV105 Sultan and a Land Rover FFR (FFR – fitted for radio) at the combat team or battle group headquarters where he would act as engineer adviser to the team or group commander and relay orders to the field squadron headquarters or field troops by radio. Similarly at a higher level the field regiment commander would be with his own tactical headquarters at the armoured division headquarters acting as an adviser to the division commander.

Each Field Engineer regiment assigned to an armoured division has at its disposal the services of a field support squadron – see accompanying diagram. This squadron is not a part of a Field Engineer regiment on paper but works very closely with one particular field regiment. Each field support squadron has three troops, a resources troop, a plant troop and a bridging troop.

The resources troop may be regarded as a stores troop, but as it deals with combat engineering materials it will hold such diverse things as timber and building materials as well as combat engineering stores. It also contains a high proportion of tradesmen like carpenters, welders and so forth whose services might be needed. The plant troop, as its name implies, holds the various items of engineering plant that the field regiment might require, over and above the holdings of the field squadrons. Some of this plant is highly specialised and among some of the more specialised plant not shown on the table are generators, circular saws, steam cleaners and such items as towed vibrating rollers, not forgetting the inevitable 2-ton maintenance binned trailers (two off) to keep check on all the various items. The table shows only the main items of plant. When these are assigned to the field squadrons they go complete with their driver/ operators who work them once on site. The bridging troop holds the assault bridging for the Field Engineer regiment and delivers it to the site where it is required. The bridging troop does not actually construct the bridge as that is left to the squadron or troop concerned. The usual allotment of bridging for each field support squadron is two medium girder bridges (MGB) and one heavy girder over bridge (HGOB). The HGOB bridging components are currently carried in AEC 10-tonne trucks but from 1986 onwards they will be replaced by a new 14-tonne truck, the Bedford TM 6-6.

Needless to say each field support squadron has its own REME section tasked with keeping all the various items of plant and the transport vehicles moving.

Each of the armoured division Field Engineer regiments can expect some support from the two Corps-controlled Field Engineer regiments. In action these two Field Engineer regiments can be expected to occupy the areas behind the armoured

Organisation and Equipment of Field Support Squadron in BAOR

Field Support Sqn HQ

Resources Tp	**Plant Tp**	**Bridging Tp**	**REME Sec**
4 x MC	1 x MC	1 MC	
3 x Land Rover	2 x Land Rover	1 x Land Rover	
5 x 4ton	2 x 4ton fuel	8 x 4ton	
1 x 4ton fuel	2 x 10ton	1 x 4ton fuel	
4 x 10ton	6 x dump truck	9 x 10ton	
3 x crane	1 x crane	(MGB)	
	1 x med excavator	(HGOB)	
	6 x MWT		
	3 x LMD		
	2 x LTT		
	6 x LWT		

Organisation and Equipment of BAOR Corps Field Plant Squadron

Plant Sqn HQ

Forward Tp

Div Det | *Div Det*

1 x MC
5 x 35ton tractor
4 x Land Rover
2 x 4ton
1 x 4ton fuel
2 x 10ton
4 x heavy excavator
2 x MTT
4 x HTT

Rear Tp

Rear Tp HQ

2 x 35ton tractor
2 x Land Rover
1 x 4ton
2 x 4ton fuel
1 x 10ton
2 x med crane
1 x dump truck
2 x med excavator
2 x med scraper
1 x med grader
2 x LMD
2 x LWT
4 x MTT

Route Maint Sec | *Route Maint Sec*

1 x MC
1 x Land Rover
3 x dump truck
1 x med grader
1 x LTT
2 x MWT

division Field Engineer regiments and provide support by keeping supply routes clear and preparing alternative defensive positions. One of these two Corps regiments has three squadrons but the other has only two plus a plant squadron – see accompanying diagram. This plant squadron is yet another example of Corps resources being held at as high a command level as practical so that they can be issued as and when required. Thus the forward Field Engineer regiments and the Corps-controlled regiments can call upon the plant held by the plant squadron.

Much of the equipment held by this Squadron is heavy and intended for route preparation, repair and maintenance rather than for the forward area combat engineer role, although much of it could be so used. As will be seen from the organisation table the plant squadron is divided into two main troops, the forward troop, with two division detachments that would be use to back up the two forward Field Engineer regiments of 1(BR) Corps, and the rear troop with its rear troop headquarters and its two route maintenance sections.

There is one support squadron left to mention, the only one of its kind in the army: 65 Corps Support Squadron (see accompanying diagram) which supports not only the Field Engineer regiments but all the other Royal Engineer combat engineer regiments in BAOR, including the Amphibious and Armoured Engineer regiments.

Organisation of 65 Corps Support Squadron

Sqn HQ

- Resources Support Troop
- Forward Delivery & Bridging Troop
- M2 Troop
- Engineer Parks
- REME Workshop

Above:
CET carrying trackway. *MoD*

The organisation table outlines the main troops of this squadron but what this table cannot indicate is the sheer size of its holdings. For instance the engineer parks contain all manner of stores worth millions of pounds; in weight alone they contain some 16,000 tons. There are in the parks no less than 1,000 tons of mines and explosives, not all of it war stock for some is always on hand for training purposes. The staff of 65 Corps Support Squadron are half civilians and half Royal Engineers, not forgetting the REME personnel who operate from their large base workshop.

In 65 Corps Support Squadron the four troops require some explanation. The Resources and Support Troop acts as a form of magnified resources troop for the field support squadrons. It holds such items as stockpiles of gravel and aggregate for construction purposes and massive stockpiles of timber for all purposes from emergency bridge-building to building construction. Just to handle all this stockpiled equipment requires a large holding of mechanical handling equipment (MHE) and a fleet of trucks to move it – because the Corps Support Squadron moves the material to where it is required. The Forward Delivery and Bridging Troop holds the war stocks of bridging of all kinds and delivers them either to a prearranged point where the Field Engineer regiments will pick them up or direct to a site where possible. The M2 Troop holds some spare M2 rigs for the Amphibious Engineer Regiment while the Engineer Parks have already been mentioned. There are three of these parks close to 65 Corps Support Squadron's base at Hameln with two more at Sennelager and Hildesheim and a small detachment at Osnabrück.

BAOR has its own repair and general support organisation for the Royal Engineers in the form of 40 Army Engineer Support Group at Willich. The main element of this is 21 Army Base Workshops while nearby is 303 Army Engineer Park and 306 Army Plant Park. These large static locations act as the repair and maintenance centre for all Corps plant and machinery and as the park names imply, they also act as a form of reserve of all equipment not actually issued to the units further forward. Again this support group contains a sizeable civilian element.

Also civilian but not really a part of the combat engineer establishment in any way are the Mobile Civilian Engineering Groups (MCEG). HQ MCEG is at Krefeld with 211 Mobile Civilian Artisan Group (MCAG) at Sennelager, 212 MCAG at Hohne, 217 MCAG at Arsbeck and 256 Mobile Civilian Plant Group (MCPG) at Hannover. These groups operate throughout the BAOR area carrying out routine repair, maintenance and construction tasks that would otherwise occupy the time of the combat units. Much of their work is routine and could be directly compared to a commercial construction concern's activities.

United Kingdom Field Regiment

There are nominally four Field Engineer regiments based in the United Kingdom although due to a division of their manpower some of the field squadrons will in time of war become airfield damage repair (ADR) units and will travel to Germany; their role will be covered in a section later in this book. This immediately reduces the total Field Engineer squadron numbers to two with most regiments and less in one case. As has already been mentioned the United Kingdom combat engineer units travel on wheels, not tracks and whereas a BAOR field section would travel into action in a FV432 the United Kingdom counterpart would travel in a 4-tonne truck or a Land Rover. They also have three field sections to a troop instead of the four in BAOR.

The Territorial Army Engineer regiments are denoted by the term (V) behind their title to indicate Volunteer. Their organisation is exactly the same as a United Kingdom Field Engineer regiment although in some cases the regiments have only two field squadrons.

As is only to be expected in a formation the size of the Royal Engineers there are organisational anomalies in the United Kingdom set-up one of which is the 69 Gurkha Independent Field Squadron which was specially formed to provide engineer support for 2 Brigade. These Gurkhas are not part of the Brigade of Gurkhas.

Then there are the two specialist units, 9 Parachute Squadron and 59 Independent Commando Squadron RE. These two squadrons are organised in a similar way to any other field squadron but as their titles imply they have very specialised roles. Despite these special roles they can and do function as ordinary field squadrons at times but their training is necessarily angled towards their method of getting into action.

The two squadrons are very similar in organisation and equipment so it will only be necessary to describe the organisation of one, 9 Parachute Squadron (see accompanying diagram). This squadron is part of 36 Engineer Regiment which supports 5 Airborne Brigade for out-of-Europe operations. The personnel are all volunteers who have to meet the exacting requirements of the Parachute Regiment as far as physique and agility are concerned and be mentally robust to face the challenge of jumping out of an aircraft to get into action – the Commando Engineers have to face the equally exacting requirements of the Royal Marines. 9 Parachute Squadron has a squadron headquarters with a headquarters troop and three troops each with about 40 officers and men. It has its own support troop with a slightly different scale from the norm, for generally speaking both the Paras and the Commandos travel light. When they jump into action they take very little of their equipment with them. That has to be landed rather than paradropped and for this reason very little of their equipment and plant has actually been modified for the airborne role. Only two items of equipment have actually undergone any alteration. One is a light wheeled tractor (LWT) that has had its cab removed to save space and the other is a light crawler tractor (LCT) that has been modified so that its dozer blade and the rear-mounted winch can both be removed to save space and make the load just para-droppable. Once they get into action the Paras and the Commandos will rely to a large extent on that ancient Sapper asset of improvisation. Once on the ground they will simply make do with what they have. Since it is expected that most operations will take place on or near airfields there is a good chance that some sort of plant will be around and the Sappers will simply use that until their own kit arrives.

The Para and the Commando squadrons are both Independent squadrons in effect so they have a slightly increased manpower level and their own REME section to keep things running. 9 Parachute Squadron will be used in support of 5 Airborne Brigade so its three troops will be divided among the three battalions of the brigade. 59 Independent

Organisation of 9 Parachute Squadron RE

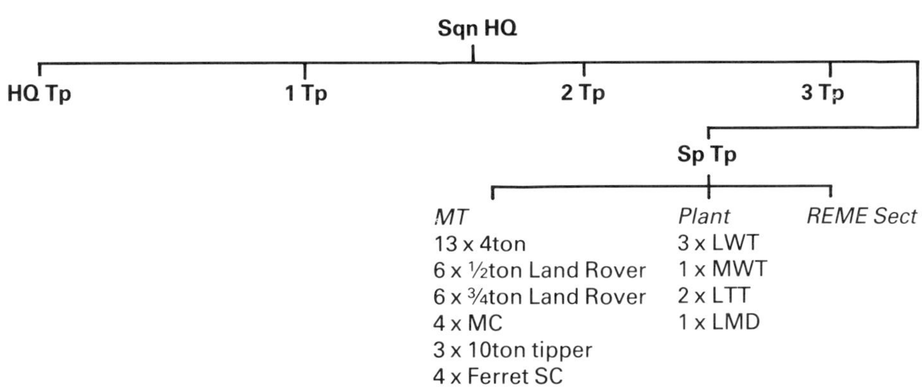

Commando Squadron RE will be used to support 3 Commando Brigade so again it can be expected that one troop will be assigned to each Commando. They will probably be supported in this role by 131 Independent Commando Squadron RE (V).

Below:
LWT with armoured cab and fork lift attachment.

Right:
Hymac tracked excavator with earth auger.

Bottom:
35ton tractor towing Craven Tasker 37-tonne RE plant-carrying semi-trailer with two MCTs. *Craven Tasker*

Combat Engineer Equipment

In the pages that follow an indication will be given of the various types of equipment that the Corps of Royal Engineers uses in the combat engineer role. It is not comprehensive as such a listing cannot include all the many and various items that are issued and used. There is space only for a very brief outline of each item of plant or equipment and some not covered in this section can be found later in the book.

Ferret Scout Car

The Royal Engineers use several marks of Ferret, mainly for reconnaissance and liaison purposes under the command of a reconnaissance sergeant. The vehicles are usually armed with a L3A3 0.30in Browning machine gun, making them the last vehicles in the British Army to be so armed. Smoke dischargers are mounted on the front hull. Some units add their own extra stowage racks on the back to provide more space internally and a fixture for the turretless Ferrets in BAOR is some form of tarpaulin cover to keep the weather out of the otherwise open hull.

The Ferrets are now getting a bit elderly – the last of them came off the production lines as far back as 1971 – but they are still a nippy and handy little vehicle. There is no sign of a replacement for some time to come.

Data: Mark 2/3
Crew: 2
Weight: 4,395kg
Length: 3.835m
Width: 1.905m
Height: 1.879m
Max speed: 93km/h

Below:
Ferret scout car with extra stowage at rear.

FV103 Spartan

The FV103 Spartan is the APC component of the Scorpion range of light reconnaissance vehicles and its main use with the Field Engineer regiments is as a troop commander's vehicle. As such it contains map boards and radios and all the other command equipment necessary. There is space in the vehicle for the driver, commander and four men although for the troop commander's use this is usually reduced. A machine gun may be mounted on the commander's cupola and smoke dischargers are provided on the front hull.

Crew: 6 with driver
Weight: 8,153kg
Length: 4.89m
Width: 2.21m
Height: 2.26m
Max speed: 85km/h

Below:
FV103 Spartan.

FV105 Sultan

The FV105 Sultan is another variant in the Scorpion light reconnaissance vehicle range and is configured specifically for the command and control role. Compared with the FV103 Spartan it is higher and has more equipment internally, including seating, map boards, extra radios and so on. Since the Sultan can expect a great deal of coming and going as people visit the command post, a special canvas penthouse can be erected on the rear to provide shelter for the 'visitors'. On the move this penthouse together with its supporting bows is stowed on the roof or the hull front. The commander's cupola can accommodate a machine gun and there are smoke dischargers on the hull front.

The Sultan is also used as a command vehicle by the Armoured and Amphibious Engineer regiments. United Kingdom Field Engineer regiments often use the Saracen wheeled command vehicle in lieu of the Sultan.

Crew: 5 to 6
Weight: 8,664kg
Length: 4.8m
Width: 2.242m
Height: 2.026m
Max speed: 72.5km/h

Above right:
FV105 Sultan with penthouse stowed on roof.

Right:
Maintenance on a FV105 Sultan.

BAY 1

FV432

The FV432 has no name but it is the British Army's standard APC. As such it is equally suitable as the vehicle for carrying into action the BAOR Field Engineer regiment's sub-unit, the field section. Each FV432 carries nine men including the corporal commander and the driver along with all the Sapper stores and equipment needed by the section and the men's personal kit and weapons.

The FV432 can be used for a variety of Sapper combat engineer tasks from simple personnel carrying to minelaying. In the latter role it can be used to tow the Bar Minelayer, with pallets of Bar mines carried inside the vehicle, or it can carry the Ranger anti-personnel mine projection system on the roof. The section crews tend to add bits and pieces to their vehicles, including extra stowage boxes on the roof or sides and sometimes large awnings over the roof hatch.

Crew: 9
Weight: 15,280kg
Length: 5.1m
Width: 2.8m
Height: 2.286m
Max speed: 52km/h

Below left:
FV432 with additional awning over roof hatch.

Below right:
Track-bashing on a Sapper FV432.

Bottom:
Engineer FV432s with their Ranger mine projectors covered.

FV180 Combat Engineer Tractor (CET)

Above:
CET with loaded bucket.

The first production CETs rolled off the production lines during 1977 and ever since then it has been one of the most versatile items of equipment in the Sapper inventory. A tracked vehicle intended primarily for earth-moving, it is also armoured and amphibious. It has a crew of two and can be driven by either crew member in either direction but on the road the 'front' is the end with the earth-moving bucket. The CET can also 'swim' with a little preparation: a floatation bag is secured in the earth-moving bucket and a trim board is fitted to the front hull. In the water the CET is propelled by two Dowty water jets. To help the CET get out of the water or over obstacles a rocket-propelled earth anchor can be fired and the CET then winches itself forward.

On land the CET can be used for many combat engineer tasks. Its earth-moving bucket can be used to construct or demolish obstacles and dig weapon or tank slots rapidly. The bucket can also be used for rough grading and by the addition of a special crane jib can lift 4,000kg. A two-speed winch is fitted that can be used to recover vehicles; it has a pulling capacity of 8,000kg. The CET can also be used to nudge or push medium girder bridges into position and the bucket can also be used to carry lengths of Class 30 or Class 60 trackway. If all this were not enough the CET can also be used to tow various trailers, including the Giant Viper trailer which it can fire in action.

The last CETs were produced during 1981 but recently India has ordered a batch and production will commence once more. It seems very likely that the Royal Engineers will then get a few more to replace those lost through wear and tear or those now in the Falklands. Two CETs gave sterling service during the Falkland Islands campaign until one was finally blown up by a mine outside Port Stanley. They were two of the few tracked vehicles that actually made the journey all the way from San Carlos to Port Stanley.

In BAOR Field Engineer regiments the CET is held by the field squadron's support troop which has a complement of five CETs. They are also used by the Armoured and Amphibious Engineer regiments.

Crew: 2
Weight: 17,010kg
Length: 7.3m
Width: 2.9m
Height: 2.667m
Max speed: 55km/h

CET in action in typical conditions.

Special crane jib in CET bucket which can lift loads up to 4,000kg.

Above:
Swimming CET with recovery anchor to left of crew hatches.

Light Mobile Digger (LMD)

The LMD is a special Sapper vehicle used to dig trenches on the battlefield. It consists of a Thornycroft Nubian 4 x 4 chassis onto the rear of which has been built an earth-digging device developed in conjunction with the coal mining industry. This cutter is raised to the horizontal and then lowered to the ground surface where teeth cut into the surface. Earth is raised to the upper part of the digger where it is dumped onto a conveyor belt which drops the spoil onto either side. The digger head is hydraulically powered and as the vehicle moves forward it can cut a trench 0.61m wide and 1.38m deep at a forward speed of from 4.57 to 5.49m/min, depending on the soil conditions. The LMD is air-portable.

Above:
LMD in travelling configuration.

Overleaf top:
Trench digging with the LMD.

Crew: 1
Weight: 9,253kg
Length(digging): 7.61m
Width: 2.31m
Height: 2.6m
Max speed: 70km/h

Tractor Wheeled Earthmoving

This is not an item of combat engineering equipment – it has been procured mainly for range maintenance and similar tasks. It is a 4 x 4 digger loader with a front-mounted multi-purpose bucket and a backhoe with a full 180-degree sweep. Powered by a Perkins A4.236 diesel developing 70hp and with a hydraulic shuttle transmission, the Tractor Wheeled Earthmoving is a MF Industrial MF50D Elite model.

Above:
Tractor Wheeled Earthmoving, the MF Industrial MF500 Elite.

Light Wheeled Tractor (LWT)

The LWT is one of the Field Engineer regiment's most useful vehicles for it is a form of wheeled bulldozer than can use its earth-moving bucket to dig or simply move earth and at the rear of the vehicle a device known as a back-acter can be used to dig trenches or weapon slots. To add to the

Top:
LWT in action as earth-mover.

Above:
The two forms of the LWT with earth bucket on left and fork lift attachment on right.

versatility the front bucket can be replaced by a forklift. The LWT is the Muir-Hill A5000 bucket loader and this model was chosen for its combination of speed, small turning circle and versatility. The Engineers have added to its versatility by developing an extra protection kit for the cab for use in locations such as Northern Ireland.

Crew: 1
Weight: 8,695kg
Length: 6.3m
Width: 2.29m
Height: 2.95m with cab
Max speed: approx 40km/h

Below:
Side view of LWT showing fork lift and back-acter.

Bottom:
LWT with cab removed as used by 9 (Para) Squadron RE.

Medium Wheeled Tractor (MWT)

The MWT is now the TEREX 72-51 which has now virtually replaced the earlier Allis Chalmers vehicle. Again it is a wheeled bulldozer but the MWT does not have the back-acter of the LWT. It does however have a powerful 11,340kg capacity front-mounted winch which can be used for vehicle recovery. The large front-mounted earth bucket has a struck capacity of 2.3sq m which means a straight static tipping load of up to 11,925kg. The first TEREX MWTs were produced during 1977 and production for the Royal Engineers continued until 1983. The MWT has the distinction of being one of the largest loads that can be fitted into a Hercules transport aircraft.

Crew: 1
Weight: 18,420kg
Length: 7.32m
Width: 2.95m
Height: 3.45m with cab
Max speed: approx 30km/h

Top:
MWT in action.

Above left:
MWT from rear showing optional routers for surface ripping.

Above right:
LWT carrying out earth-clearing operations.

Light Tracked Tractor (LTT)

The LTT is actually a bucket loader that can be fitted with a number of accessories including a back-acter for trench digging. It is a military version of the International Harvester 100B.

Crew: 1
Weight: 8,074kg
Length: 4.78m with winch
Width: 1.9m
Height: 2.6m with cab
Max speed: 7.4km/h

Left:
LTT on 10-tonne tilt trailer.

Below left:
Sapper conference around a LTT.

Above:
LTT with back-acter belonging to 39 Engineer Regiment (EOD).

Above:
MTT in its element on the Chattenden Ponderosa.

Medium Tracked Tractor (MTT)

The MTT currently in use is the Caterpillar D6D although earlier models such as the D6C are still in service. Both are commercial models with some modifications to enable them to meet military standards. The MTT may have surface-ripping routers mounted at the rear. The cab can be removed.

Crew: 1
Weight: 17,790kg
Length: 2.84m
Width: 3.4m with blade
Height: 3.1m
Max speed: 10km/h

Above:
MTTs in an engineer park in BAOR – these are Caterpillar D6Cs.

Below:
Loading a MTT onto a Craven Tasker 37-tonne RE plant-carrying semi-trailer. *Craven Tasker*

Heavy Tracked Tractor (HTT)

The large HTT is currently the TEREX 82-30B, the first of which was procured during 1979 and a further batch in 1981. The HTT is a powerful machine with a large hydraulically-powered dozer blade and is very popular with its driver/operators because it is one of the few British Army vehicles to have an air-conditioned cab.

Crew: 1
Weight: 32,020kg
Length: 7.01m with blade and winch
Width: 3.376m
3.78m with blade
Height: 3.78m with cab
Max speed: 10km/h

Above:
The TEREX 82-30B, the current HTT.

Dump Truck

Two types of dump truck are currently to be found in service with the Royal Engineers. The oldest model is the Scammell LD55 with a heaped capacity of 9.1cm but this is now being replaced by the Haulmatic 615 with a maximum payload of 13,180kg. The engine of the Haulmatic is a Perkins V-8 diesel of 8.831 litres, and unlike the earlier Scammell vehicle it has a one-man cab.

Data: Haulmatic
Crew: 1
Weight: (loaded) 24,640kg
Length: 6.6m
Width: 2.3m
Height: 2.9m
Max speed: 55km/h

Above:
Scammell LD55 dump truck.

Above:
The Haulmatic 615 dump truck.

Self-loading Dump Truck

This dump truck is one of the latest pieces of equipment to be procured for the Royal Engineers as it was ordered in late 1984. The vehicle is a 6 × 6 Scammell S26 powered by a Rolls-Royce Eagle 350 Mk 3 turbocharged diesel. The dump body has a capacity of 9.8cu m and when not being used as a dump truck it can be used to carry general cargo or up to seven standard NATO pallets. Behind the cab is an Atlas hydraulic self-loading crane that can be fitted with a hook or a 0.5cm grab. The vehicle is designed for use at temperatures down to −35°C and it can restart on a 56% gradient.

Crew: 1 + 1
Weight: (loaded) 32,520kg
Length: 8.86m
Width: 2.5m
Height: 3.5m
Wheelbase: 5m
Max speed: 79km/h

Below:
The Scammell S26 self-loading dump truck. *Scammell*

Wheeled and Tracked Excavators

The Royal Engineers use several types of excavator, wheeled and tracked. The wheeled version is the Medium Excavator, the International Harvester 630W; the tracked equivalent is the Hymac 590.

Data: IH 630W
Crew: 1
Weight: approx. 14,600kg
Length: 6.79m travelling
Width: 2.5m
Height: 4.67m travelling
Max speed: 40km/h

Above right:
Hymac tracked excavator.
Right:
Medium excvator, the IH 630W.
Far right:
Medium excavator with arm extended.

HY-MAC

Medium Grader

The current Medium Grader in use by the Royal Engineers is an Aveling Barford model, the Super MGH with a 12ft/3.66m mouldboard for the actual surface grading.

Crew: 1
Weight: approx 13,380kg
Length: 7.85m
Width: 2.5m
Height: 3.3m
Max speed: 43km/h

Above:
Aveling Barford medium grader.

Below:
Aveling Barford medium grader. *MoD*

Motor Scrapers

The Royal Engineers use two types of motor scraper, the TEREX TS-14B described as a Heavy Scraper and the TEREX TS-8 described as a Medium Scraper and light enough to be loaded into a Hercules transport aircraft. Both are used mainly by the BAOR Field Plant Squadron.

Data: TS-14B
Crew: 1
Weight: (loaded) 47,480kg
Length: 13.22m
Width: 3.45m
Height: 3.5m
Max speed: 37km/h

Above right:
TEREX TS-8 medium scraper.

Below right:
Craven Tasker 37-tonne RE plant-carrying semi-trailer being used to carry medium scraper. *Craven Tasker*

TEREX

Below:
Front view of the TEREX TS-14B heavy scraper.

Cranes

The Royal Engineers use several types of mobile crane for many tasks ranging from bridging to stores handling. For many years one of the standard field cranes has been the Coles Hydra Husky 150T and many of these are still in service with the Corps. In recent years these have now been joined by a new Crane, Field, Medium, the Coles 315M. The Coles 315M can lift a load of 15 tonnes at a radius of 3m.

Data: Coles 315M
Crew: 1 + 1
Weight: 21,130kg
Length: 9.14m
Width: 2.5m
Height: 3.6m
Max speed: 75km/h

Above:
The Coles Hydra Husky 150T.

Below:
The new Crane, Field, Medium, the Coles 315M.

FV622 Truck, Cargo, High Mobility Load Carrier Stalwart Mk 2

The Royal Engineers is not the only corps in the British Army to use the Stalwart but they are used in some numbers by the Engineers. The Stalwart is an amphibious high mobility load carrier with a 6 × 6 configuration that can tackle just about any type of rough terrain while carrying a load of up to five tonnes, so it is a very useful vehicle indeed. It is

Right:
Royal Engineer FV622 Stalwart Mk 2s.

issued down to troop level and every troop headquarters has one in all BAOR Field Engineer regiments. The Sapper Stalwarts usually lack the hydraulic crane of the FV623 variant. First produced in 1966, the last example was issued in 1971 so the Stalwart is now getting to be a rather elderly vehicle. There is no sign of a replacement because of cost, and so recently a major refurbishing programme was completed to keep the vehicle in service for some years to come.

Crew: 1 + 2
Weight: loaded 14,480kg
Length: 6.356m
Width: 2.616m
Height: 2.54m to cab
Max speed: 63km/h

Below:
FV622 Stalwart Mk 2 with canvas tilt covering cargo-carrying area.

4tonne GS Truck

The Bedford 4tonne truck is one of the most widely used of all British Army vehicles and it is in service in many forms. The current model is the Bedford MK although the earlier Bedford RL will still be encountered, especially in the United Kingdom. The model usually used by the Corps is the General Service or (GS) version with a canvas tilt covering the cargo area which has drop sides and tailgate. However the Corps does use the truck in a flat-bed form to carry large pallets or crates; one special version in this form is the 4tonne Bulk Fuel which carries a dismountable fuel tank and dispensing unit to keep other vehicles topped up in the field. These fuel trucks are usually operated by the support troops in every field squadron.

The 4tonne trucks are also used to carry Medium Girder Bridge kit components and can be converted to carry and lay lengths of Class 30 trackway. A snow plough may also be fitted. Many trucks used by the Corps have extra stowage added over the cabs. A tipper variant is also in service.

Crew: 1 + 1
Weight: loaded 9,590kg
Length: 6.35m
Width: 2.489m
Height: 3.4m
Max speed: 77km/h

Right:
Royal Engineers 4tonne truck with extra stowage over cab.

Above:
4tonne truck bulk refuelling.

Below:
Class 30 trackway being laid from frame carried on 4tonne truck.

Above:
4tonne truck about to cross a Medium Girder Bridge. *MoD*

Below:
4tonne dump truck used by 1 Fortress STRE on Gibraltar.

10ton Truck

The 10ton truck is a type that is now destined to leave the Army's ranks in the near future and the Royal Engineers is now the largest user of the type. The most common type of 10ton truck used by the Corps is the AEC Mk 1, a 6 × 6 truck with a payload of slightly over 10 tonnes. As such it is a perfect heavy load carrier for much of the Corps heavy equipment such as the Heavy Girder Overbridge. The Corps also has a number of Leyland 10ton trucks still in service.

Below:
10ton truck with hydraulic crane used by 28 Amphibious Engineer Regiment.

Bottom:
10ton truck with canvas tilt.

Both types are now due to be phased out of service as soon as the new Bedford TM 6-6 14ton truck comes into service after about 1986. It will not be too soon for the REME fitters for both the AEC and Leyland vehicles are now getting rather long in the tooth and are well overdue for replacement. Some of them have already been replaced by the new Bedford TM 4-4 8ton trucks in a few units.

Crew: 1 + 1
Weight: loaded 19.100kg
Length: 9m
Width: 2.49m
Height: 3.58m
Max speed: 50km/h

Below:
AEC Mark 1 10ton truck.

Truck 10ton 6 × 4 GS (Albion) Machinery RE

The vehicles known usually as the Machi trucks, or something similar, are among the Corps' oldest and are deserving of special mention. With the designation FV11102 they entered service during the early to mid-1950s and have been around ever since, lovingly cared for and a reminder of past days. These Albion trucks are among the very last of their kind in service and are used as mobile workshops where the Corps craftsmen can turn their hand to making and repairing all manner of things. The interior of the box body contains machine tools, tool cabinets, work benches and all the necessary lighting and heating. They are usually to be found in support squadrons although some have been replaced or supplemented by 4-tonne trucks with workshop bodies.

Crew: 1 + ?
Weight: 15,740kg
Length: 8.28m
Width: 2.515m
Height: 3.454m
Max speed: 43.4km/h

Below:
Truck 10ton 6 x 4 GS (Albion) Machinery RE.

35ton Tractor

The 35ton tractor is used by the Royal Engineers for the carrying of large items of plant such as MCTs and HCTs although many other similar loads can be carried. The vehicle currently in use in the Scammell Crusader 6 × 4 tractor truck that can tow a variety of semi-trailers. The 35ton tractor is fitted with a winch to assist in loading and off-loading and in an emergency this winch can be used for self-recovery.

The 35ton truck may be fitted with some special semi-trailers. One is the Craven Tasker 37-tonne RE plant-carrying semi-trailer which has a 'goose neck' that can be lowered from the tractor end to allow loading after which it is raised again for towing. A special trailer is used to tow the No 9 bridge for the Chieftain AVLB but perhaps the rarest semi-trailers used with this tractor truck are the special M2-carrying semi-trailers used for M2 recovery after breakdown.

Incidentally these vehicles may be seen with both 'Scammell' and 'Leyland' on the front.

Crew: 1 + 3
Weight: empty 11,095kg
Length: 6.66m
Width: 2.5m
Height: 3.3m
Max speed: 65km/h

Below left:
Scammell Crusader 35ton tractor carrying a Blaw-Knox Paver (PF 90).

Below right:
35ton tractor carrying Coles Hydra Husky 150T crane.

Bottom:
35ton tractor towing semi-trailer with FV432.

Unimog

It would be difficult to describe this little truck as a piece of combat equipment but it is essential for use in BAOR park and depot areas to keep access routes clear at all times. Its main role is as a snow clearer for which purpose snow clearing attachments, including ploughs, are fixed ot the front of the vehicle; the same attachment points can be used for sweepers. The model is actually a Unimog U 600L.

Crew: 1 + 1
Weight: 2,250kg
Length: 4.74m
Width: 1.825m
Height: 2.25m
Max speed: 73km/h

Below:
One of the Unimogs used to keep depot and park access routes clear of snow in BAOR.

Bar Mine and Bar Minelayer

The Bar Mine is now a standard item of combat engineer kit and has been for some years. It is an anti-tank mine made almost entirely of plastic to make detection very difficult and its length of 1.2m increases the chances of a tank running over at least part of it once it has been laid. These mines are layed in large minefields and which can be sown quite quickly by means of the Bar Minelayer, a plough-type device towed behind a vehicle, usually a FV432, and through which the mines are fed by hand for the actual laying. The plough blade on the end of the layer it so arranged that it opens up a furrow, places the mine inside it and then places the earth over the mine for concealment. Using this layer it is possible to lay between 600 and 700 Bar Mines in an hour but this requires a constant re-supply. The mines are usually supplied in pallets holding 72 mines and they are automatically fuzed as they pass through the layer. 4tonne trucks can be used for the laying operation or vehicles such as the Stalwart or 1-tonne Land

Right:
FV432s towing Bar minelayers and carrying Ranger mine projectors on parade.

Rover but the lighter vehicles lack the traction to allow the plough to create a furrow and thus lay the mines on the surface. Using the Centurion AVRE mines can be laid from a towed trailer.

Recently the Bar Mine has been enhanced by the addition of new types of fuze. A double impulse fuze to defeat mine-clearing rollers is one recent modification but two other fuzes have been added to the list of options to convert the Bar Mine into what is known as the Full Width Attack Mine (FWAM), referring to the full width of the target tank underside. One of these new fuzes uses a mast sensor that moves against the tank belly while the other is an electronic fuze. These new FWAM fuzes can be used with the Bar Minelayer.

Each troop of a field squadron has at least one Bar Minelayer.

Mine
Weight: 11kg
Length: 1.2m
Weight: of explosive 8.4kg

Bar Minelayer
Weight: 1,240kg
Length: 4.2m
Width: 1.6m
Height: 1.27m

Right:
Bar minelayer.

Anti-tank Mine Mk 7

The Anti-tank Mine Mk 7 is very much a relic of a previous generation of mine warfare as it uses a metal case for the main charge and is thus fairly easy to detect using metallic mine detectors. It has to be laid by the time-honoured shovel-and-hand method but despite this it is still a powerful anti-tank weapon and its service life has been extended by the use of a frangible carbon rod mast sensor. Normally the mine is set off by a weight of about 275kg on the top pressure plate but the new carbon rod system sets off the mine when the rod is broken by a tank or other vehicle passing over it to release a spring-loaded mechanism to detonate the mine.

Weight: 14.7kg
Diameter: 325mm
f28Height: 130mm
Weight: 8.89kg main charge (TNT)

Above:
Anti-tank Mine Mk 7 with carbon sensor rod.

Mine, Anti-tank, Non-metallic L3A1

This anti-tank mine has a rubber case and has only a pressure plate mechanism operated by a weight of between 120 and 200kg. It is armed by turning a simple lever device on the pressure plate and to disarm the mine the lever is simply returned to the safe position. This mine is not a standard issue item and is kept mainly for training and as a reserve for use by home defence units.

Weight: 7.7kg
Diameter: 266mm
Height: 145mm
Weight: 6kg of explosive (TNT)

Above:
Mine, Anti-tank, Non-metallic L3A1.

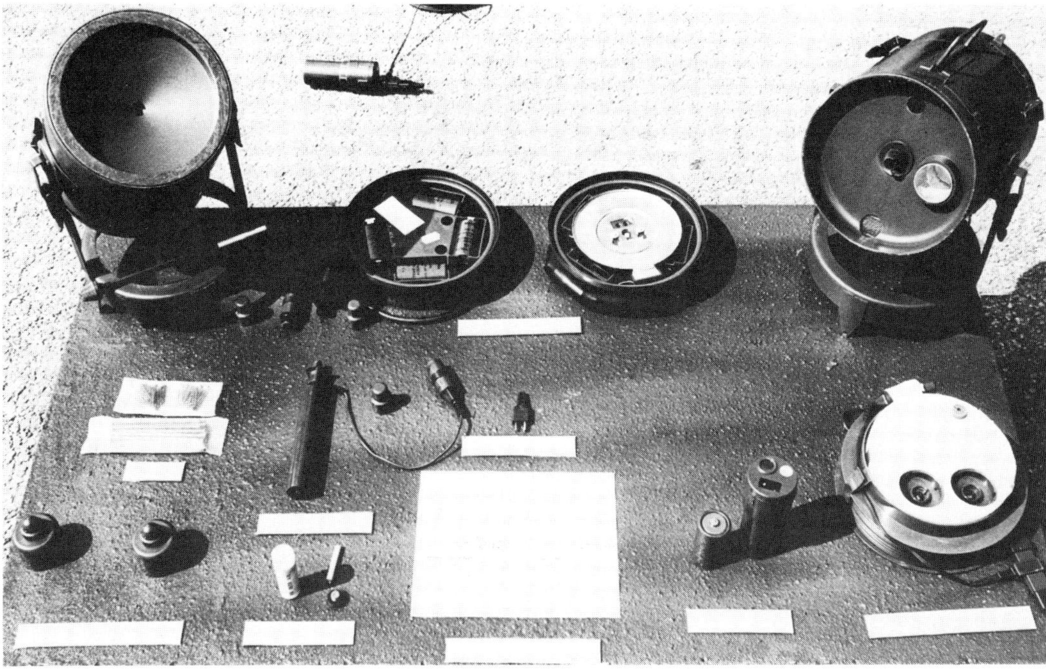

Off-route Anti-tank Mine (ORATM)

The ORATM was originally a French design known as the MIACAH and although the mine bodies were imported from France in the first place they were filled with explosive in the United Kingdom. The idea behind the ORATM is that it operates horizontally across a route likely to be used by enemy tanks or vehicles. When a tank crosses a tape sensor or wire across the road the mine fires and projects a hollow charge at the tank to

Above:
ORATM complete with all accessories. *GIAT*

Below:
The new Lawmine that will replace the existing ORATM.

penetrate the side armour. The mine is effective up to 80m and it can be aimed with a high degree of accuracy.

In the foreseeable future the ORATM may be replaced by a new device known as the LAWmine or Lawmine. This uses the projectile from the LAW80 anti-tank rocket weapon. The rocket with a diameter of 94mm is contained in a short barrel next to a sensor that fires the rocket when a target is in range.

Weight: 13kg
Height: 330mm
Body diameter: 185mm

Ranger

Ranger is an anti-personnel mine system intended for the rapid sowing of small mines to delay the advance of an enemy. Its tactical uses are many and vary from the rapid scattering of anti-personnel mines across small rivers to cover approach routes to a crossing to 'seeding' an anti-tank minefield to delay clearance.

The ranger mine is circular, 62mm in diameter and 32mm deep containing approximately 10gm of RDX/Wax which is enough to maim but rarely sufficient to kill. These mines are packed into disposable tubes in magazines. Each tube holds 18 mines and a magazine hold s 72 mines so each magazine has four tubes. The magazines are loaded into a discharger which holds 18 magazines; thus a fully loaded discharger holds no less than 1,296 mines and these are fired from their tubes in a pre-programmed sequence. Each tube is fired using a small black powder charge which projects it s 18-mine contents to a maximum range of 140m. The projector can be adjusted in elevation and traverse to suit the required range and the density of the minefield. Once the mines land there is a short delay before they are primed and then a light pressure will cause them to detonate. For training purposes a mine made from peat is used; obviously these do not need clearing after use!

Ranger projectors are usually carried on FV432s but they can be used from other vehicles such as the Combat Support Boat (to cover river approaches) and Stalwart.

Weight: 647kg discharger-loaded
Range: 140m
Length: 2.28m
Width: 2.2m
Height: 1.3m
Number of mines: 1,296

Below:
Sapper loading magazines of Ranger mines into discharger.

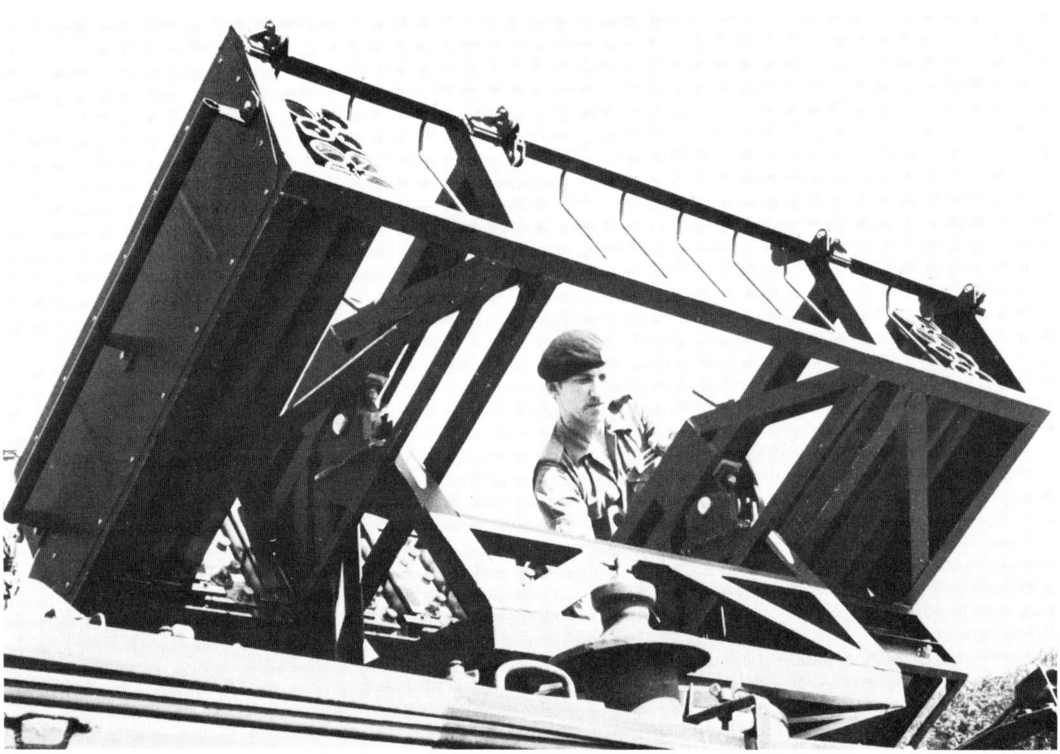

Below:
FV432 carrying loaded Ranger discharger.

Mine Detector No 4C

The Mine Detector No 4C is but one of several types of mine detector in use by the Royal Engineers. Certainly one of the oldest, it is still a reliable and efficient metal detector. It uses two electrical coils that create an electrical field which is distorted by the presence of metal. When this disturbance occurs a signal is produced in the user's headphones. The detector may be used on the end of a long handle but a much shorter handle can be used when the detector is being used from the prone position.

Weight: 14.4kg in case
1.8kg head
Handle length: 1.28m extended

Below:
Mine Detector No 4C in use.

Demolition Materials

The Royal Engineers are the Army's demolition experts and are equipped accordingly. One of their most widely used explosives is a substance known as PE4, a plastic explosive that looks exactly like plasticine – indeed, plasticine is used for training purposes because PE4 is moulded and handled in exactly the same way. PE4 is perfectly safe to handle and can only be detonated by a detonator. Another specialised plastic explosive is Sheet Explosive No 2 that can be cut to any shape simply by using scissors! Charge Linear Cutting (CLC) also makes use of plastic explosive but this time it is encased in lengths of lead sheath to produce a hollow charge effect. Lengths of CLC can be cut to suit almost any cutting demolition job and various sizes are available. The largest size can cut through 15mm of mild steel, and it can be bent and shaped to cut holes through metal plates if necessary.

Various specialised demolition stores are available. One is known as a 'beehive' charge but is officially the Charge Demolition No 1 6-in Mk 6. This is a shaped TNT charge used to punch a hole through concrete or steel. Another charge used for the demolition of bridges or other steel structures is the Charge Demolition Necklace L1A1, sometimes known as the Hayrick charge from its general shape. Actually the L1A1 is a kit with a number of charges and attachment links and clamps – even nails are provided in case they are needed.

The Sappers also make use of cratering kits to make roads impassable to vehicles. These are usually two-part kits that use one charge to make a small hole in a surface into which a much larger charge is placed to produce a much larger crater. Using a standard kit craters up to 8m wide and 3m deep can be produced.

Above right:
Various sizes of Charge Linear Cutting (CLC). *Royal Ordnance*

Left:
Charge Demolition No 1 6in Mk 6.

Above left:
The No 14 charge, part of the Charge Demolition Necklace L1A1.

Right:
Explosive Kit Rapid Cratering. *Royal Ordnance*

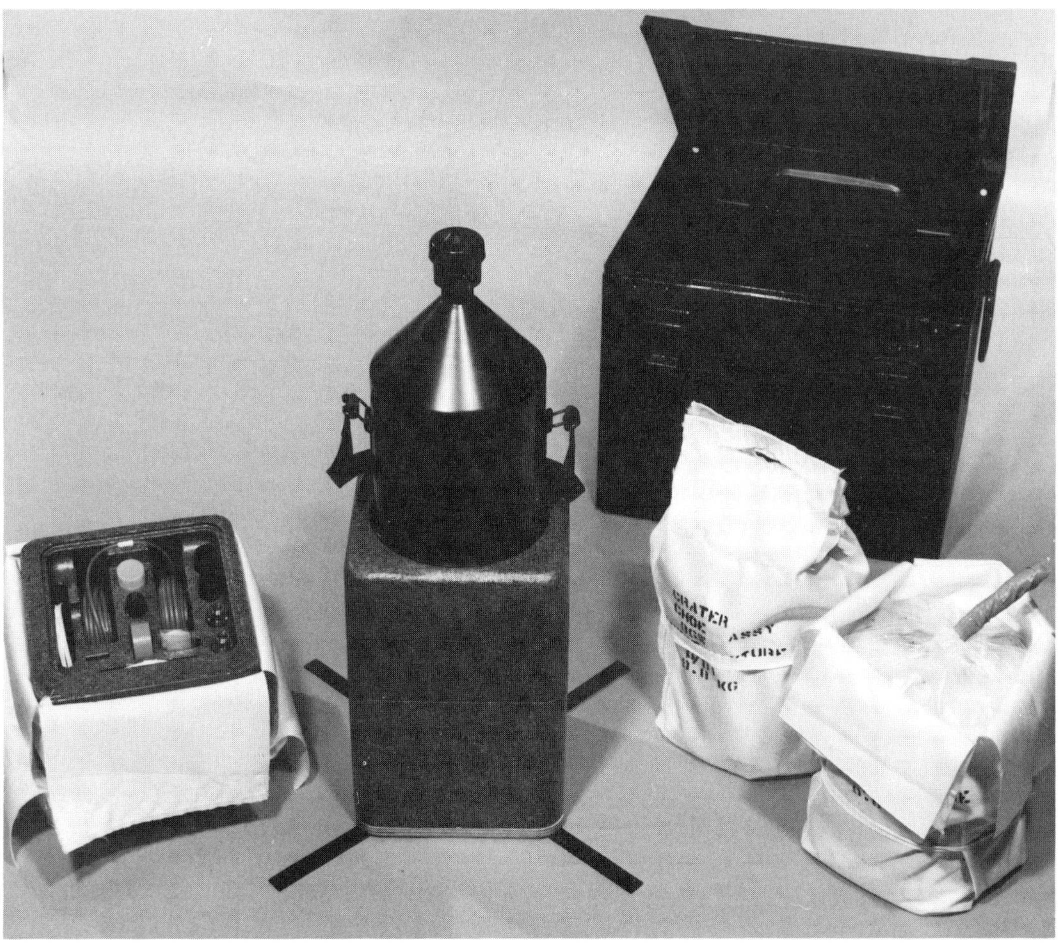
CRATER
ASSY
KG

Medium Girder Bridge (MGB)

The MGB is one of the Royal Engineer's most valuable pieces of equipment and used widely for a variety of support bridging applications. The MGB can be constructed over wet or dry gaps of up to 30m to carry loads up to Class 70, ie strong enough to carry a Challenger tank. If longer bridges are required a pier set is available to support the extra spans and by using a link reinforcing set bridges up to 45m long can be built still capable of carrying a Class 70 load.

A 30m bridge set is carried on five 4-tonne trucks towing special trailers. By using more sets longer bridges can be constructed. The MGB sets are held by the Field Support squadrons who deliver the bridge sets to the site where they are to be constructed by the Field Engineer squadrons. Once on site the bridges are constructed by hand. The individual components are made from an aluminium alloy and there are only seven main components in all. All the components can be manhandled into position and once in place are secured by bolts. The actual construction can be reduced to a drill and this drill is one of the most frequently practised routines carried out in training by the field squadrons. A practised crew can erect a Class 70 bridge in a surprisingly short time but actual figures are hard to quote due to the various bridge lengths and local conditions involved. Demonstrations have been given where 30m Class 70 bridges have been constructed in about 45min.

The actual construction for a 'standard' bridge starts with the placing of a building frame at the point to be crossed. A nose section is then erected and bridge sections are then constructed. As each section is ready the constructed sections are gradually nudged across the gap by a 4tonne truck, a CET or any other handy vehicle, until the nose reaches the opposite side of the gap where a

Below:
Abbott 105mm self-propelled gun crossing a MGB. *Fairey Engineering*

Right:
A Chieftain tank crossing a floating MGB Class 60 bridge. *Fairey Engineering*

Below right:
MGB with link reinforcing set to enable longer bridges to take Class 70 loads; bridges up to 45m long can be constructed using this set. *Fairey Engineering*

roller is placed to take the load – spare bridge sections can be used if necessary to counterbalance the load during this operation. Once across more sections are constructed until the bridge is complete. Ramp sections are then added. The link reinforcement set, if used, is added as the bridge is built. If the portable pier set is to be used this can be slung from the underside of the bridge until it is launched to the correct location. Pontoons may be used to support a bridge on occasion and the same pontoons can be used to support bridge sections to produce a floating ferry.

The MGB is a very versatile piece of kit and the sets are manufactured by Fairey Engineering Ltd. New uses and parts for the bridge are still being developed and it will remain in service with the Royal Engineers for many years to come.

Below:
Standard MGB in position. *Fairey Engineering*

Bottom:
Sappers of 68 Gurkha Field Squadron finishing off a MGB in Hong Kong. *MoD*

Heavy Girder Overbridge (HGOB)

The HGOB is not intended to construct bridges but is used to strengthen existing culverts or damaged bridges to allow loads of up to Class 60 to cross. If necessary the HGOB may be laid across an existing bridge or culvert to allow tanks and heavy equipment to pass without problems. The maximum span that can be bridged is 7.6m.

Airportable Bridge (APB)

The APB is not a standard item used by BAOR as it can only carry a Class 16 load but it is a very useful piece of engineer kit. Intended primarily for limited war operations the APB can be assembled by hand and when assembled is light enough to be carried into position by a Chinook helicopter. When in kit form the APB is carried on Land Rovers pulling trailers and a complete set requires 20 special trailers. The set can be used for several purposes – as it is light enough to float it can be used to make four rafts, each 22m long. Alternatively it can be used to make 64m of floating bridging, but its basic use is to construct a bridge 15m long and 4m wide to cross a dry gap. When used as a raft up to four outboard motors are used to propel the finished product and such a raft can be constructed by 24 men within 45min. A dry gap bridge can be constructed within 20min by a crew of 16 men.

Above:
APB being used as a floating raft to carry a 4tonne truck.

Above:
CSB at speed.

Combat Support Boat (CSB)

The CSB is one of the British Army's success stories during recent years for although it was originally developed to assist the building of various types of military bridge, it has become a far more versatile little vessel. This was proved during the Falkland Islands campaign when the CSB was used as a carrier for all manner of loads from men to supplies and was strong enough to deliver its loads right over the beaches.

The first CSBs were obtained during 1980 and they are produced by Fairey Allday Marine Ltd. Powered by Dowty water-jets that provide the craft with a speed of 30km/h even when fully laden, the use of the water-jets enables the CSB to operate in very shallow water. They also enable it to brake from full speed ahead to a dead stop within the length of the craft. The CSB has a small cockpit forward and later production versions provide extra heating and general provision for operating at low temperatures. The area behind the cockpit can be used to carry all manner of stores including the Ranger mine projector for the seeding of river banks with anti-personnel mines. But the CSB was intended for assisting bridge construction and thus the bows have large pushing 'knees' for nudging pontoons and bridge components into position.

The CSB has been exported to the US Army for bridge construction purposes, and in sizeable numbers. With the Royal Engineers it is carried on special launching trailers. The cockpit housing can be removed to make the load lower.

Weight: 4,080kg
Length: 8.2m
Beam: 2.5m
Height: 2.79m
Max speed: 40km/h unladen

Below:
CET being used to assist launching of CSB from its special trailer. *Fairey Marine*

Bottom:
CET without bow 'knee' as used by 28 Amphibious Engineer Regiment.

Below:
Front view of CSB showing 'knees' for nudging bridging components.

3 Armoured Engineers

Armoured Engineers serve in 32 Armoured Engineer Regiment (the 'Armoured Farmers') which can trace its origins back to the 79th Armoured Division formed during World War 2 – although it has been disbanded and reformed since then. Much reduced in size since WW2, the Regiment now carries out many of the operational tasks performed by the 'Assault Engineers' of the war years. The Regiment was reformed in 1980 and is now based at Munsterlager. It consists of a headquarters, three Armoured Engineer squadrons (26, 31 and 77), a REME workshop and an RAOC stores section. The RHQ is split into a tactical HQ and a regiment echelon. Each of the three Armoured Engineer squadrons (see accompanying diagram) is affiliated to an armoured division. Each squadron has three armoured troops, equipped with a total of nine AVLBs and nine AVREs, supported by a squadron echelon. Plans for future regimental expansion allow for the formation of up to an additional troop in each squadron.

The individual squadron headquarters are based on two Sultan command vehicles, with a further Sultan providing the HQ for the squadron echelon. The echelon is controlled by the squadron admin officer and consists of a stores section run by the SQMS, an MT/support troop and a fitter section. The MT/support troop is equipped with four Crusader tractor and trailer units, 24 B vehicles of various types, three CETs, a LWT and a Coles Hydra Husky Crane.

The armoured troop has a troop commander mounted in a Spartan, with a troop staff sergeant and a recce sergeant mounted in Ferret scout cars. This small troop HQ controls three Centurion AVREs and three Chieftain AVLBs. Troops tend to have specific brigade and battle group affiliations,

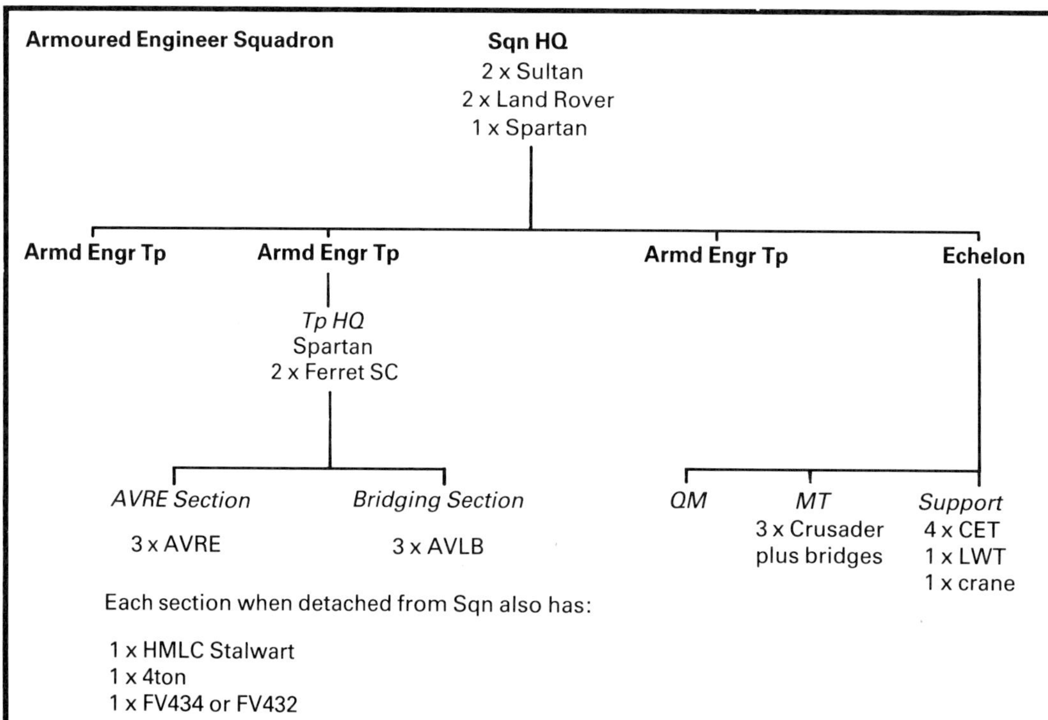

often working independently of their SHQs. When this occurs the squadron's support and echelon elements can be split so that each troop is allocated a CET, a 4-tonne stores-carrying truck and a REME FV434 repair vehicle.

The current role of the Armoured Engineers is the same as it has always been: to maintain the mobility of our own forces and deny mobility to the enemy. The Armoured Engineers are essentially Combat Engineers trained to operate under armour, with specialist Engineer tanks. They can keep up with the main battle tanks and can carry out engineer tasks both at speed and under fire.

The primary tasks carried out by the Armoured Engineers are:

(a) The rapid creation of obstacles.
(b) The breaching of minefields.
(c) The breaching/crossing of obstacles, both natural and artificial.
(d) The destruction of enemy strongpoints.

Specific tasks depend on the nature, phase and type of war being waged. In the advance, Armoured Engineers would carry out route and obstacle clearance, rapid bridging, and track construction. During the assault, they would be responsible for bridging and minefield breaching forward of the start line, the rapid clearance of scatterable mines, obstacles and routes and the clearing of strongpoints. In the defence and withdrawal, Armoured Engineers would assist in rapid anti-tank ditching, the digging of positions, demolitions and, to supplement divisional obstacle plans, the laying of mines and the construction of minor obstacles. Additional tasks that Armoured Engineers might be called upon to assist in are route maintenance in battle group areas, deception plans and collective NBC measures.

The bridging role of the AVLB is often demonstrated in times of peace. Its ability to provide assault bridging across wet and dry gaps is readily appreciated and understood. The role of the AVRE however, is not so obvious in peacetime, but in a conflict it would prove invaluable.

The AVRE is the workhorse of the Armoured Engineers and it is continually being adapted to use new equipments as they come into service. It can be fitted with a dozer blade that can be used for anything from clearing debris from routes to digging 'tank slots'. It can carry and lay Class 60 trackway and fascines, the trackway being used to

Below:
AVRE 165 with turret bustle moved to hull rear for fascines or trackway to be fitted; the turret is still traversed to the front.

provide a firm approach to and exit from gap crossing sites, whilst the fascines provide a quick and easy method for tracked vehicles to cross ditches and small gaps. The AVRE can be fitted with a mine plough (as can Chieftain AVLBs) which can be used in conjunction with Giant Viper for minefield breaching operations. AVREs are equipped with a demolition gun which can be used against a variety of targets ranging from bridge abutments and pillboxes through to area targets such as power stations. The AVRE's towing hook allows it to tow a 7½ton cargo trailer or a Giant Viper trailer.

Changes in tactical doctrine in BAOR have placed greater emphasis on more mobile tactics and the necessity for fast moving, hard striking counter move forces. This doctrine has re-emphasised the need for Armoured Engineers and has led to the Regiment's current expansion programme. This has meant that there is a requirement for more tracked vehicles, both AVRE and AVLB. Because the present AVRE was based on the Mark 5 165mm Centurion gun tank which has not been produced for many years, the additional AVREs have had to be found from elsewhere. Centurion gun tanks, previously used by the Royal Artillery's Forward Observation Officers, have been converted and have been designated as AVRE 105s to be used primarily as 'plough tanks'. The second troop in each Armoured Engineer squadron has become an AVRE 105 troop, thus giving each squadron a mine plough capability. In addition it has also been necessary to find additional AVLBs. These are being obtained from the conversion of a number of early marks of Chieftain MBT, to produce a new standard Mk 6 Chieftain AVLB, the first of which was due in service by the end of 1985.

Other new equipments recently brought into service with the Armoured Engineers include the new 'Maxi Pipe Fascine' which replaces the old Chesspale fascine. The fascine is designed to be carried on a special cradle mounted on the front of the AVRE from which it can be dropped into an anti-tank ditch so as to give a rapid means of crossing for AFVs. The Maxi Pipe Fascine is basically an outer band of 220mm diameter plastic pipes held together with chains and stuffed with

Left:
Chieftain AVLB crossing a No 8 bridge and carrying a No 9 bridge with reconnaissance dinghy on top.

loose pipes which are retained by a net at either end of the fascine. It all weighs approximately 3 tonnes as compared to the Chesspale's 11 tonnes, its loose construction enabling it to conform to the shape of ditches thus ensuring a more level crossing.

Each Armoured Engineer squadron is also equipped with a number of CETs. These amphibious armoured vehicles are designed specifically to provide integral engineer support to the battle group. In gap crossing operations they are primarily used to prepare and maintain approach routes, this includes the laying of trackway and pipe fascines. They can also be used to tow and fire Giant Viper.

The 35 tonne Crusader tractor and trailer units, which are held by the armoured squadron's MT/support troops, are designed to transport the bridges for the AVLB. These bridges are the 23m span No 8 folding scissor bridge, and the 12m span No 9 tank bridge. The Crusader is capable of loading a No 9 bridge onto the ground using its own equipment. It is used mainly to re-supply spare bridges to the forward battlefield areas, ground dumping them ready for collection by the AVLBs. The Crusader's two-man crew is also trained to strip, build and service the bridges.

Much of 32 Armoured Engineers Regiment's initial driver training and AVRE gunnery training is carried out at the RE Wing of the RAC Centre at Bovington in Dorset. This is also the location for the annual Armoured Engineer Troop Commanders' Course. Additional training takes place within the Regiment and this is reflected in the fact that regimental personnel tend to stay for 'longer than normal' tours, often returning to the Regiment later in their Service careers. In this way the specialised Armoured Engineer knowledge and expertise is kept within the Regiment.

The Regiment has an extremely busy annual training programme. It starts with detailed crew training and tests in the winter, progressing onto troop and squadron exercises, and culminating in bridge camp and a number of other FTXs towards the end of the year. The regiment supports virtually every battle group on training exercises on the Soltau Training Area, and it is the only unit of any arm which has personnel on every battle group exercise at BATUS in Canada. As such, the Regiment is one of the most active Engineer regiments in BAOR.

Armoured Engineer Equipment

Centurion AVRE

The first Centurion AVRE was a Centurion Mk 5 tank converted to the Armoured Engineer role during 1957. It was the early 1960s before the first examples were issued for service in the regiment and it has remained in service ever since. Although no replacement is due in the foreseeable future the present tanks have a vital role to play in BAOR and are constantly adapted to use new equipments, maintained to a high standard and kept in readiness at all times.

The main feature of the Centurion AVRE 165 is the 165mm demolition gun mounted in the turret. This powerful weapon is accurate and effective up to a maximum of 2,400m but provides its best performance at ranges up to 1,000m. It fires a 29kg HESH projectile which is powerful enough to bring down most hard targets such as buildings and bridges of concrete and masonry. The 105mm gun of the AVRE 105 is not as powerful as the 165mm but it too fires a HESH projectile weighing 11.26kg.

The L7A1 gun of the AVRE 105 can fire armour-piercing ammunition but operationally it will be used to fire HESH for demolition purposes only. The current secondary armament of the AVRE is two 0.30in in L3A3 Browning machine guns, one coaxial with the main gun and the other mounted over the commander's roof hatch. In the future these will be replaced by a form of 7.62mm GPMG to make them compatible with other army machine guns.

Apart from the armament the AVRE has a front-mounted dozer blade but as yet the AVRE 105s do not have this fitted; they have mine ploughs instead. Above the dozer blade is a cradle for carrying fascines and the same cradle can be used

Below:
AVRE 105 with mine plough raised and towing Giant Viper.

Bottom:
Bovington-based AVRE 165 used for training.

Top:
AVRE 105 on laser range for gunner training – note laser projector over gun barrel.

Above:
Training on the use of Class 60 trackway carried on a AVRE 165.

to carry Class 60 trackway. Using explosive blow out pins the AVRE can lay the fascine and trackway direct from the cradle in a closed down position. The towing hook at the rear can be used to tow a special trailer (see next entry) or a Giant Viper.

The Centurion AVRE has a crew of five: commander, gunner, radio operator/loader, driver and co-driver. The driver has a difficult task as the Centurion has a really heavy clutch control pedal but when the cradle above the dozer blade is occupied by a fascine or length of Class 60 Trackway his forward vision is severely restricted. The turret bustle at the rear of the turret is also used to hold items such as kit and camouflage netting but has to be removed and re-located to the top of the engine decking when fascines or trackway is carried as this requires the gun to be traversed to the rear and the bustle would then interfere with the load.

Data: AVRE 165
Crew: 5
Weight: 50,394kg unladen
Length: 8.686m
Width: 3.39m hull
3.96m with dozer blade
Height: 3.009m
Max road speed: 34.6km/h
Armament: 1 x 165mm demolition gun (105mm gun on AVRE 105) 2 x 0.30mm or 7.62mm machine guns.

Below:
Breech of 165 mm demolition gun.

AVRE Trailer (FV2721)

The AVRE trailer was designed especially for use with the Centurion AVRE and once it has been connected to the AVRE it can go anywhere the AVRE will tow it. Its carrying capacity is 7,500kg and in use this weight can be made up from anything including fuel, ammunition, engineer stores and so on. It can also be used to tow a Bar Minelayer and pallets of Bar Mines are then carried on the trailer to be fed down into the minelayer as the AVRE moves forward.

Capacity: 7,500kg
Weight: 7,316kg empty
Length: 6.59m
Width: 2.972m
height: 2.083m

Below:
AVRE trailer on tow behind 'racing' AVRE 165 with side skirts and dozer blade removed.

Bottom:
AVRE trailer loaded with an early form of Maxi-Pipe fascine.

Giant Viper (GV)

Giant Viper is a mine clearing equipment consisting of a hose 250m long filled with plastic explosive. The hose is carried in a special trailer and projected across a minefield by a cluster of solid fuel rocket motors. When the hose is in place the explosive in the hose is detonated and the resultant hefty blast can clear up to 75% of all blast susceptible mines in a lane approximately 200m long and 8m wide.

The Giant Viper trailer (FV3705) is usually towed by an AVRE but the CET and even the FV432 can also be used. The crate holding the hose weighs 2,136kg and is loaded into the trailer using a forklift truck. A new type of trailer introduced recently is a converted heavy floating bridge (HFB) trailer with two close-coupled wheels each side. This holds the hose crate on a flat-bed platform.

Data: FV3705

Weight: 4,180kg loaded

Length: 5.95m

Width: 2.28m

Height: 1.83m

Below:
FV432 towing Giant Viper trailer at BATUS, Canada.

Right:
Giant Viper rocket motors.

Below:
New type of Giant Viper trailer using a converted HFB (Heavy Floating Bridge, now obsolete) trailer as the basis.

Mine Plough

Ordered during 1982 the Mine Plough is a relatively new piece of kit for the Armoured Engineers. It consists of two ploughshares carried in front of each track in such a way that the plough and its seven tines digs down into the ground, digs out any mines in the vehicle's path and moves

Below:
Mine plough fitted to AVRE 165.

Above:
One of the mine ploughshares showing the seven tines.

them over to one side. A chain between the two ploughshares is used to trigger any mast-sensor anti-tank mines. Mines down to a depth of about 230mm can be cleared. Each plough share clears an overall width of 3.767m with a central unploughed gap of one metre. Shoving the mine plough through the ground takes a great deal of effort from the user vehicle. The Centurion AVRE has to operate in bottom gear and forward speed is approximately 4km/hr.

The Mine Plough is carried by the AVRE 165 and 105 and may also be used by the Chieftain AVLB to increase its tactical flexibility.
Weight: 2,300kg
Overall clearance: 3.787m
Uncleared centre-lane width: 1m

Chieftain AVLB

The Chieftain AVLB was first delivered to the British Army in 1974 following development that went back to 1962. With 32 Armoured Engineer Regiment the AVLB actually consists of three elements. One is the Chieftain bridgelayer, the other two are the bridges, the No 8 and No 9 versions. The bridges are carried on a Crusader tractor and trailer.

On the Chieftain AVLB the bridges are emplaced using a series of hydraulic rams under the control of the driver. The bridgelayer is a turretless Chieftain with various hydraulic rams and cylinders along with a bridge support system. Of the two bridges used, the No 8 is a folding scissors-type assault bridge with a maximum span of 22.9m. The No 9 bridge is a one piece bridge with a maximum span of 12.2m. The Crusader tractor has the ability to lay a No 9 bridge directly from the trailer over existing civilian low classification bridges.

When a wet gap is to be crossed using an AVLB a careful reconnaissance is always carried out if conditions and time permit. The usual procedure is to send a small two-man diving team to carry out a reconnaissance of the river width and profile. All this is carefully recorded for the next planning stage. If the gap can be crossed with a single bridge, all well and good, but if the gap is too wide (over 23m) for a single bridge span, combination bridging has to be used. One bridge is laid initially with its end onto the bed of the gap. A second AVLB is then driven onto this bridge and launched so that the second bridge rests upon the first. The launching of the second and subsequent bridges must be done with precision. It is a procedure which has to be practised frequently even in the

Above:
Multi-span bridge in the process of laying using one No 8 bridge as the laying platform for another.

Below:
Chieftain AVLB with No 8 bridge.

worst weathers and in the dead of night. The most difficult part of the whole operation is the removal of bridges after use. The bridgelaying boom has to engage in a lug on the bridge and getting the bridgelayer and boom into the correct position for engaging is a delicate and difficult exercise. The Mk 6 AVLBs will, when they are issued, have an intercom on the bridge boom to enable the commander to guide the driver.

The new Chieftain Mk 6 AVLBs will be about 3,000kg heavier than the existing versions and the first of them will be in service with the Regiment by 1986.

Crew: 3
Weight: 53,300kg with No 8 bridge
Length: 13.741m with No 8 bridge
Width: 4.165m
Height: 3.923m
Max road speed: 48km/h
Armament: 2 x 7.62mm machine guns

Data: No 8 bridge	**Data**: No 9 bridge
Weight: 12,700kg	**Weight**: 8,840kg
Length: 24.384m	**Length**: 13.411m
Width: 4.165m	**Width**: 4.165m
Max span: 22.9m	**Max span**: 12.2m

Above:
No 8 bridge in the process of laying.

Below:
Bovington-based Chieftain AVLB in action.

Above:
Crusader Tractor with No 9 bridge on special semi-trailer.

Below:
Depth-sounding from a bridge reconnaissance dinghy.

4 Amphibious Engineers

The Amphibious Engineer Regiment – 28 – nicknamed the 'Rubber Ducks', operates under the control of 1(BR) Corps and provides support to all units within 1(BR) Corps by providing rapid bridging across rivers and canals in the Corps area. It was formed in 1970 and is based at the pleasant town of Hameln (Hamelin), using the River Weser as its training area.

Of all the many Royal Engineer regiments 28 has perhaps the most complex organisational structure (see accompanying diagram) and it also has one of the largest base workshops of its kind set up by REME to look after the Regiment's equipment.

The equipment that makes the Regiment amphibious is the M2 bridging and ferry system, a West German system based on floating wheeled vehicles and developed during the 1960s. The M2 is also used by 28's equivalents in the West German Bundeswehr and co-operation between 28 and the West Germans is close. The M2 is basically a large wheeled vehicle that can float and propel itself in water. By using a system of ramps and flat decks, numbers of M2 units, known to their crews as 'rigs', can be joined together to form 'instant bridges' and ferries. This form of rapid bridging will be very necessary in time of war for bridges will be among the very first communication points to be either deliberately demolished or rendered useless by enemy action and as the subsequent holding and counter-strike actions unfold there will be many urgent requirements to get men and vehicles across water obstacles. The importance of this rapid bridging can be seen in the fact that a single 1(BR) Corps armoured brigade has approximately 1,000 vehicles. Using M2 bridges about 120 vehicles an hour can cross most of the 1(BR) Corps area river and water obstacles so even using two M2 bridges it will still take between four to six hours for a single armoured brigade to move across. Obviously the more rapid bridges that can be provided the better. In an emergency the River Weser would be a prime candidate for 28's attentions and the Aller, Leiner, Elbe and the mighty Rhine are others, not forgetting the many major canals that criss-cross Germany.

To carry out this bridging role 28 Amphibious Engineer Regiment has a regimental headquarters that forms its own tactical headquarters if the regiment ever had to go to war. The two main M2 users are 23 and 64 Amphibious Engineer Squadrons and they are supported by 71 Amphibious Engineer Support Squadron and the considerable resources of 28 Amphibious Engineer Regiment Workshop (REME). Each of the

Outline Organisation of 28 Amphibious Engineer Regiment

RHQ

Tac HQ
2 x Sultan
(2 x Ferret SC (War only))
7 x Land Rover
1 x 4ton
1 x MC

- **23 Amphibious Engineer Sqn**
  - 1 Tp
  - 2 Tp
- **64 Amphibious Engineer Sqn**
  - 3 Tp
  - 4 Tp
- **71 Amphibious Engineer Support Sqn**
- **28 Amphibious Engineer Regiment Workshop (REME)**

Above:
Challenger MBT crossing bridge formed from M2Ds on River Weser. *RSME*

two Amphibious Engineer squadrons (see accompanying diagram) has two troops but to complicate matters the regiment always maintains a number of M2 rigs under wraps during peacetime. In war these rigs would be ready for immediate use and they would be added to those maintained by the two squadrons. Thus in peace each squadron has two amphibious troops each with 12 M2s. In war this would be increased by an extra three rigs for each troop, making 15 rigs to each troop. Additional spare rigs are held by 71 Amphibious Engineer Support Squadron who have as one of their components a reserve rig troop with three sections each holding three M2s.

There is also the Amphibious Training Troop. In normal circumstances the Regiment carries out all its own routine training that deals with the operation and running of the rigs by using this 'in house' troop. As with the other amphibious troops in the two squadrons this training troop has the four peacetime amphibious sections that would expand to five in an emergency. The entire Amphibious Training Troop would then become a fully operational amphibious troop and would go off to war with the rest.

A third component of the Amphibious Engineer Support Squadron is the Amphibious Support Troop. This has three sections, one being the Diving Section that has ready to hand trained divers to carry out underwater reconnaissance of likely bridging points, and for all the other missions that an amphibious unit will require. There is also a Plant Section with six CETs for the preparation of bridge approaches and so forth; there are also two LWTs in this section. The third section is the Resources Section which has ready to hand four CSBs and their trailers.

The actual vehicle strengths of the squadrons can be seen in the accompanying diagrams. In an emergency the entire regiment would probably leave its base at Hameln and move out into carefully selected 'hides' out in the country close to likely bridging points. There the amphibious troops would be joined by the REME Forward Repair teams who would allocate their repair and

Organisation of Amphibious Engineer Squadron

Amph Engr Sqn HQ

1 x Sultan
1 x Ferret SC (+ additional Ferret SC in war)
5 x Land Rover
2 x 4ton
1 x 10ton (to be replaced by 8ton)
1 x 4ton fuel
1 x MC

Amph Tp

Tp HQ

1 x Spartan
3 x Ferret SC
1 x Land Rover
2 x 4ton
1 x 14ton (fitted with crane)
1 x MC

Amph Sec — Sec — Sec — Sec — Sec* — Wksp Sec

Each: 3 x M2
*Note: not manned during peace

Amph Tp

Wksp Pl
(in the field)

recovery capabilities to the potential sites and 'hides'. Some additional assistance in moving the Regiment's equipment could be expected by the allocation from elsewhere of heavy trucks and tippers for carrying ballast and other material to improve bridge approach tracks and roads.

The amphibious sections with their rigs would be taken to carefully selected positions to await their call. When that call is made the preliminaries should already have been made. The Diving Section will have made its underwater reconnaissance and the CETs will have carried out their preparation work on both banks. These two operations are but two that will have been carried out before the rigs lumber forth from their field sites. With help from the Royal Military Police the rigs will move in a set order to the bridging site and if all is ready they will enter the water. As the sections enter the water they manoeuvre into their pre-selected bridge configurations. If the bridge is to be used for heavy tanks they will align themselves closely side-by-side and latch together. If only light vehicles are to use the bridge they can space themselves out more and connect themselves by using their ramps; which would be fitted into place using the on-board cranes. These operations would continue until the water obstacle is completely bridged and the ramps at each end are in place. Anchorages have to be secured at each end of the bridge which would be used to help the M2 engines maintain the bridge in position and the traffic can then cross. The CSBs will be on hand to act as safety boats ready for any emergencies.

The amount of traffic crossing the bridge at any one time will depend on many factors. The length of the bridge will be one factor and so will the actual weight of the vehicles involved. Heavy vehicles such as tanks usually cross one at a time while vehicles like Land Rovers can cross in strings. The introduction of the Challenger to BAOR has caused a few problems. The M2 was originally intended for loads up to Military Load Class 60, ie 60 tonnes, but the Challenger exceeds this limit. Thus the M2s have had to undergo a programme of modification to take this extra load, a programme which has involved the installa tion of extra buoyancy bags to the sides and front of each rig.

Always close to hand will be the REME Forward

Below left:
Preparing an M2 for the road after training.

Organisation of 71 Amphibious Engineer Support Squadron and its relationships with the Regiment in the field

RHQ

- **64 Amph Engr Sqn**
- **23 Amph Engr Sqn**
- **71 Amph Engr Sp Sqn**
  - *SHQ*
  - 1 x Ferret SC
  - 5 x Land Rover
  - 2 x 4ton
  - 1 x 10ton (to be replaced by 8ton)
  - 1 x 4ton fuel
  - 1 x MC
- **A2 Ech**
  - (Coloc with Amph Engr Sp Sqn)
  - 1 x Land Rover
  - 3 x 4ton
  - 1 x 10ton (to be replaced by 8ton)
  - 2 x 8ton bulk fuel
  - 1 x MC
- **Wksp Main**
- **B Ech**
  - (Rear Party)
  - 1 x staff car
  - 1 x car utility
  - 1 x Land Rover
  - 3 x 4ton
  - 1 x MC

Under 71 Amph Engr Sp Sqn:

- *Amph Trg Tp*
  - Tp HQ
  - 4 x Ferret SC
  - 1 x Land Rover
  - 2 x 4ton
  - 1 x 14ton (fitted with crane)
  - 1 x MC
  - Sec, Sec, Sec, Sec, Sec*
  - Each: 3 x M2
- *Amph Supp Tp*
  - Diving Sec: 1 x 4ton
  - Plant Sec: 1 x Ferret SC, 6 x CET, 2 x LWT
  - Resources Sec: 1 x Land Rover, 5 x 14ton (fitted with crane), 4 x CSB + Trlrs, 1 x MC
- *Reserve Rigs Tp*
  - Tp HQ
  - 2 x Ferret SC
  - 1 x 4ton
  - 1 x MC
  - Sec: 3 x M2
  - Sec: 3 x M2
  - Sec: 3 x M2
- *Wksp Pl* (in the field)

*Note: not manned during peace

Repair teams. Using mainly Land Rovers they will be ready to carry out any immediate repairs but any major rectifications will have to be carried out at Workshop Main. For this purpose the REME teams have at their disposal two Scammell 35ton tractors pulling semi-trailers. A broken-down M2 cannot be towed so it has to be carried on these special semi-trailers.

The REME's involvement with 28 Amphibious Engineer Regiment is large and as an example of how the REME is integrated into the Royal Engineers' operations the Workshop's outline organisation is now examined (see accompany table). The Officer Commanding the Workshop has under him a large factory-like workshop devoted to the repair and maintenance of the M2 rigs. In the field the Workshop splits into Workshop Main and three Workshop platoons, two with each of the amphibious squadrons and one with the Amphibious Support Squadron. The two squadron platoons each have two amphibious troop forward repair sections while the support squadron has one section for the amphibious training troop. When these platoons operate with the amphibious squadrons they come under their control but when working with the Workshop Main they remain under the command of the REME OC.

The M2s take some looking after. At any one time a number of rigs from the Regiment are with the original manufacturers undergoing what may be regarded as a virtual rebuild. Repairs and maintenance at the purpose-built Workshop Main can be quite extensive but much of the M2 construction uses a form of stressed skin construction using aluminium. Thus the workshop has to train its own non-ferrous approved welders (including Sappers) and other craftsmen to deal with the various forms of heat-treating of various components.

Most of the training for the Regiment is carried out by the Amphibious Training Troop of 71 Amphibious Engineer Support Squadron. All Sappers who join the Regiment to work on the M2s have to go through a basic 2½ week course involving the operation of a rig and all the skills required. After that a crewman may have the choice of either water or land duties with the regiment. After approximately 4 months as a Crewman he can take a further four-week course to qualify him either as a rig Pilot or as a rig Driver. M2 rig drivers first have to be qualified HGV drivers. There is also an M2 commander's course for NCOs and officers that deals with all aspects of bridge reconnaissance and design.

Amphibious Engineer Equipment

M2

The M2 was designed and developed during the mid-1960s and the first was delivered to the West German Army in 1968. The British Army got theirs in 1969. M2 were produced by a consortium of two manufacturers, Eisenwerke Kaiserslautern (EWK) and Klöckner-Humboldt-Deutz (KHD), both of which actually produced full vehicles (the

Outline Organistion of 28 Amphibious Engineer Regiment Workshop

OC

- **Workshop Main**
  - **OC Wksp Main**
    - *Prod Pl*
      - Gen Rep Sec
      - Veh Rep Sec
    - *HQ & Admin Pl*
    - *MT, REC & RAOC Pl*
      - MT/REC Sec
      - RAOC Stores
- **Amph Sqn Wksp Pl**
  - *Amph Tp Fwd* Rep Sec
  - *Amph Tp Fwd* Rep Sec
  - *Ech Sec*
- **Amph Sqn Wksp Pl**
- **Amph Supp Wksp Pl**
  - *Amph Trg Tp* Fwd Rep Sec
  - *Ech Sec*

Vehicle complement of Workshop

6 x recovery vehicles
2 x 35ton Scammels with trailers
21 x Land Rover
24 x 4ton
7 x 8ton
3 x 10ton

Above:
An M2 in its purpose-built 'hangar' at Hamelin.

manufacturer of any particular rig can be identified by the fact that its badge is on the left of the pair mounted below the cab windows). These two companies are still concerned with the M2 for every seven years each rig is returned to them for a complete overhaul and virtual rebuild. About 14 rigs every year go through this process.

The current operational model of the M2 is the M2D. The original M2A was used for trials only while the M2B was the main operational model used by the British Army. The M2B is now being modified to M2D standard which has a Class 70 loading for tracked vehicles – thus enabling it to carry a Challenger – and Class 93 for wheeled vehicles. The extra buoyancy is provided by inflatable bags along the sides and modified hydraulics to cope with the extra weight. Inflated by two small compressors fitted to the rig, the buoyancy bags have non-slip panels for the crew. To complete the picture the M2C was a diesel-engined version sold to Singapore and the M2E was an interim vehicle with a new diesel engine and a hydraulic crane for the ramps in place of the current manual model. The M2E is in a state of limbo at the moment pending a decision on a vehicle known as the M3. This will be an overall improvement on the M2 with a less-stressed construction, a new engine and only three ramps to each side in place of the current four. It will have increased buoyancy to carry the new heavy loads but as yet it is still a long way off and awaiting funding.

On the road the M2 travelling width is cut down by folding the two side-mounted buoyancy tanks up and over the main body. Just before entering the water these tanks are moved hydraulically into position and locked. Once the rig has entered the water the side ramps are swung into position by the manually-operated crane and the various rigs join together in whatever bridge or ferry configuration is required. The M2 has a crew of four, including the pilot. Two engines provide the power for land travel and marine propulsion. Only one of the engines (normally starboard) is used on land but on water the port engine normally drives the centre propeller while the starboard engine normally drives the side propellers. The centre propeller can be rotated through 360 degrees and is used to steer the rig; the side propellers are fixed.

A section of three M2s can form a ferry to carry a Chieftain or Challenger tank although the normal method of using the M2s is to form a bridge. M2s have been used to bridge the Rhine.

Crew: 4
Weight: 22,000kg
Length: 11.35m
Width: 3m road; 5.92m water
Height: 3.7m (with rotating light for road travel 3.92m)
Max speed: 60km/h (road)
8kt (single rig with no payload in water)
6kt (3-rig ferry with 60ton tank in water)

Above:
M2s on the Weser with their extended ramps giving them the appearance of water-bugs.

Below:
M2 with two ramps mounted.

Above:
Operating the M2's ramp crane.

Below:
An M2 leaving the Weser.

Above:
One of the two special M2-carrying semi-trailers.

Below:
An amphibious section of three M2s forming a ferry. *MoD*

5 Airfield Damage Repair

1965 was a significant year for the Royal Engineers because it was in that year that they took over the task of construction engineering support from the Airfield Construction Branch of the Royal Air Force. Two years later came the Arab-Israeli Six-Day War when the Israeli Air Force knocked out the Egyptian Air Force by a pre-emptive strike that created so much runway damage that any aircraft which survived the first strikes were unable to take off. That lesson was well noted by the Royal Air Force and it has led to a major reorganisation of Royal Engineer responsibilities and to a major investment in plant and resources.

The immediate result of this reorganisation was the formation of 39 Engineer Regiment with four squadrons known as construction squadrons – see accompanying diagram. It has also led to the formation of a number of Territorial Army squadrons formed specifically to provide airfield damage repair (ADR) support for Royal Air Force bases in the United Kingdom. In preparation for the worst, dumps of aggregate and other materials have been made on many Royal Air force airfields, and plant has been pre-placed ready for immediate use. This investment has been substantial for 39 Engineer Regiment alone has equipment valued at over £30million.

The four field squadrons (construction) are Nos 48, 50, 52 and 53. Each of them has three field troops plus a support troop and for further support they have the services of 60 Field Support Squadron. These four squadrons have the task of providing ADR for the airfields of Royal Air Force Germany and that entails not just runway repair (or repair of aircraft operating surfaces – RAOS) but the maintenance and repair of electrical and mechanical facilities and the repair and maintenance of fuel and water supplies. They are also able to assist in the provision of ground defences.

The four squadrons are normally based in the United Kingdom but in an emergency would move to the four 'clutch' airfields of RAF Germany where one squadron would go to each airfield. Only then would 39 Engineer Regiment command all four squadrons – in time of peace it has only two field squadrons (construction) under its direct control. These are 48 and 53, normally based at

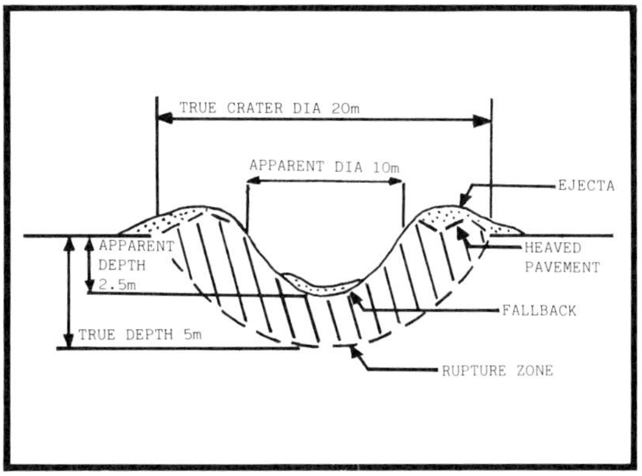

Above:
Drawing of a cross-section of a NATO standard crater.

Waterbeach in Cambridgeshire along with 34 Field Squadron and 60 Field Support Squadron. The other two squadrons are under the control of other engineer regiments. 50 Field Squadron (Construction) is under the control of 36 Engineer Regiment at Maidstone and 52 Field Squadron (Construction) is at Tidworth under the command of 22 Engineer Regiment. However, Waterbeach is now considered to be the 'home' of ADR although every year all four squadrons do travel to Germany to train on their allotted airfields with their pre-positioned equipment and plant. Co-operation with the Royal Air Force is very close.

Thus 39 Engineer Regiment only comes together as a formation in an emergency. In the process the non-ADR field squadron (34 Field Squadron) would move to support 19 Infantry Brigade but the all-important REME 39 Engineer Regiment Workshop would remain. They would be joined by four teams from the Military Works Force with specialised bulk petroleum and other fuel-handling skills and knowledge and on each airfield the field squadron would take over from the Public Services Agency (PSA) the running of the airfield's day-to-day facilities and services. Thus the OC of a field squadron (construction) would be responsible not only for his squadron but the STRE(BP) – Specialist Team Royal Engineers (Bulk Petroleum) – and the Engineer Works Organisation (EWO) for the airfield.

Organisation of a Field Squadron (Construction)

HQ Troop

1 x Saracen
4 x ¾ton Land Rover
3 x 4ton
2 x ¾ton trailer (tlr)
1 x 2ton Office tlr

Field Troop | **Field Troop** | **Field Troop** | **Support Troop**

Field Troop

1 Saracen
1 x ½ton Land Rover
1 x ¾ton Land Rover
2 x 4ton
1 x ¾ton tlr
1 x 1¾ton tlr
1 x ¾ton gen set
1 x 10cum/min compressor tlr
1 x 6ton sweeper rot airfd
1 x ½ton gen set 6kVA
6 x dump truck med
2 x excavator wh hy
1 x grader mot
1 x compactor ADR 5ton
1 x 5ton tlr AP
1 x MTT
2 x MWT
2 x HWT
1 x pump tlr
6 x mixing machine scab repair
1 x BDRM carrier
2 x lighting tower tlr
1 x towed screed beam

Support Troop

2 x tractor and Stlr 35ton
1 x ¾ton Land Rover
2 x 4ton
1x 4ton flat with machine shop
2 x 4ton fuel
1 x ¾ton tlr
2 x 2ton maint binned tlr
3 x 10ton tilt tlr
1 x ¾ton circ saw
1 x 1ton cleaner steam
1 x 1ton gen set 24/16kW
1 x ½ton gen set 40kW
1 x 2ton lub & svc
3 x dump truck med
3 x LWT
2 x MWT
2 x mixing machine scab repair
2 x vibrator screed board
1 x towed vibrating roller
1 x hydraulic hammer

The prime responsibility for the ADR field squadrons is of course airfield damage repair but they can also function as normal field squadrons and frequently do so in times of peace. But once they move to Germany their prime role is to keep the airfields functioning and for this a number of set procedures have been established. The organisation for this has already been mentioned above but exactly what would happen after an enemy attack on one of the 'clutch' airfields?

It is anticipated that any enemy air attack would take the form of a low-level strike using rockets, bomblet or scatterable minelet distribution devices and the well-established 'iron' bomb. The use of various forms of chemical or biological weapons cannot be ruled out so the men of the field squadrons (construction) are trained to work wearing full NBC kit, as are all other Sapper units.

The exact extent of the damage following the strike is one of the first things to be established and thus a reconnaissance of the airfield will be made. This will be by using either a Gazelle light helicopter or one of the squadron's Saracens. The Royal Air Force teams will also make a reconnaissance in one of their Scimitar light tanks for they have the responsibility of clearing any unexploded bombs or other devices from the airfield.

The reports from these teams will be made to the airfield Combat Operations Centre (COC) and the extent of the damage assessed. Using a large scale map the squadron commander will then sort out his various options to render the airfield operational and one of his main considerations will be the provision of a minimum operating strip (MOS) measuring a nominal 1,500m long and 15m wide to allow the aircraft based on the airfield to operate.

It is anticipated that the airfield runways will be liberally pitted with bomb craters and these will have to be eliminated before aircraft can operate. A crater is defined as a cavity which extends below the lower surface of the concrete forming a runway or taxi-way. To provide the squadron commander and his planning team with some form of yardstick a thoughtful NATO has established the form and shape of a typical crater. It has an apparent diameter of 10m and an apparent depth of 2.5m. The word apparent is used because the bomb explosion will cause damage beyond these limits to a diameter and depth of twice these figures. Some of the materials from the crater will be distributed around its rim (ejecta) and some will fall back into the hole (fallback).

Using this information the squadron commander will work out how to position the MOS bearing in mind such factors as landing approach paths and the minimum amount of work to be carried out. Once the MOS has been established the RAOS teams will move into action. Each squadron will have three RAOS teams consisting of an officer and 39 Sappers and they will move into action along well-rehearsed lines using a form of 'drill' to deal with each crater.

With this drill the two most used vehicles will be the two Michigan 275 heavy wheeled tractors (HWT). These massive vehicles are used by the ADR squadrons only and their carrying capacity is central to the crater-filling operation. Their first task, together with the team MWTs, will be to clear ejecta from the crater rims. A Grader will commence to clear debris from the general area while a MCT will clear the ejecta material by spreading it at a distance. Excavators will then start to clear the heaved concrete and the Arrow Hydraulic Hammer will cut away damaged concrete. Away at the dumps of material the dump trucks will load with 45-65mm single size aggregate to back fill the crater to within 500mm of the surface. The final 500mm is completed by compacting, in two layers, loads of 11-16mm stone with the final surface levelled using a long hollow steel scraper known as a screed beam.

The time is then ready for the repair mat which is carried from its pre-positioned dump by the HWT. This mat is a portion of Class 60 trackway measuring some 16m wide and once in position next to the repaired crater it is rolled out by hand using virtually the whole team. The HWT then gives the mat a final tug to stretch it as far as possible to prevent any aircraft passing over it causing 'ripples' to occur. The ends of the mat are then fitted with ramps and it is bolted into position using Kangol drills. Finally the whole area is brushed clear of all debris, both by hand and by the use of a Lacre rotary sweeper. The process is completed once the Royal Air Force have inspected the finished mat and aircraft can then operate.

The whole operation is co-ordinated by the use of hand-held radios in the plant cabs. Practice and training make the process as smooth as possible.

Craters are not the only method of damaging runways for cannon fire can be equally damaging to a surface by chopping out small areas that are then known as scabs. These will have to be filled in by two-man teams using Land Rovers with a mini cement mixer. Using a magnesium phosphate cement with a liquid activator these scabs can be filled in as few as two minutes, which is just as well for there is likely to be many of them; for instance at the Port Stanley airfield in the Falklands there were well over one thousand.

The runways will not be the only ADR activity after an attack. Fuel installations are likely to take a pounding and the ADR team will be in action to re-lay pipes under the advice of the STRE(BP) team. New electrical cables will be laid and damaged items repaired using special rapid-repair kits and water pipes will also be attended to. The boiler houses and generating stations will be under the control of ADR personnel and so will the many other services essential to the operation of the airfield.

Left:
The starting point, a typical bomb crater. *MoD*

Above:
MWTs start to move ejecta from the rim of the crater. *MoD*

Below:
Dump trucks filling the crater with aggregate. *MoD*

Below:
The crater has the last top surface of in-fill before the compactor moves in. *MoD*

In the United Kingdom RAOS will be provided by the TA field squadrons of 12 Engineer Brigade (ADR). This Brigade is headquartered at Waterbeach and has under its command four Territorial Army squadrons, 216, 218, 219 and 277 Field Squadron (ADR) (V) as well as 39 Engineer Regiment. These squadrons are smaller than their Regular equivalents as their role is confined to RAOS only. The first of them was 277 Field Squadron (ADR) (V), formed at RAF Leuchars in 1979. In time it is hoped that 12 Engineer Brigade will field eight ADR(V) field squadrons.

10 Field Squadron

Although 10 Field Squadron is not part of the ADR organisation it has close ties with the Royal Air Force as it supplies part of the Royal Engineer support for the Royal Air Germany Harrier squadrons, Nos 3 and 4 Squadrons, RAF. 10 Field Squadron normally lives close to the Harrier squadrons at their base at RAF Gütersloh but operationally they would move out with the Harriers into the field. On mobilisation the squadron is joined by the headquarters and two field squadrons of 38 Engineer Regiment from the United Kingdom and this Royal Engineer force will provide support in the form of landing mats, taxiways, communication routes and all the other requirements of the Harriers. The Sappers prepare the field sites for the aircraft and maintain the sites for as long as the Harriers remain in the field.

Airfield Damage Repair Equipment

Heavy Wheeled Tractor (HWT)

The Michigan 275B HWT was obtained specifically for the ADR squadrons. This large vehicle is powered by a 380hp engine and is fitted with a 5cu m capacity bucket. However it is used for other tasks than the clearing of ejecta with the bucket for it is used to tow a screed beam, carry the repair mat to its appointed position and tug the positioned repair mat once it has been rolled out. Two are used by every ADR field troop.

Below:
An HWT pulling a screed beam over the filled crater. *MoD*

Above:
Class 60 trackway mat about to be lifted from a trailer by a HWT. *MoD*

Left:
All available hands rolling out the repair mat. *MoD*

Below left:
Kangol drill being used to bolt down the edge of a repair mat. *MoD*

Below:
The last stage; brushing clear any small items of debris. *MoD*

Above right:
Another aspect of ADR; repairing damaged electrical cables. *MoD*

Below right:
A Field Squadron (Construction) with its equipment. *MoD*

Above:
The huge bucket of the HWT.

Below:
A HWT about to lift a screed beam preparatory to towing it across a compacted crater in-fill. *MoD*

Above:
An Arrow hydraulic hammer used to cut away sections of damaged concrete from around the edges of a crater.

Below left:
A Coles Model 335M dynamic compactor, used to compact the final in-fill. *Coles Cranes*

Right:
The undramatic but essential Lacre rotary sweeper used to clear debris from the area of a repaired crater prior to its use by aircraft.

Below:
A repair mat being towed into position by a HWT still on its special Dyson trailer. *MoD*

6 Explosive Ordnance Disposal

The Royal Engineers assumed the role of bomb disposal in 1940 and have retained it ever since; the title changed to Explosive Ordnance Disposal (EOD) in 1969. The main responsibility of the Corps is the aircraft-delivered explosive device but as will be related, the regiment in question – 33 Engineer Regiment (EOD) – also deals with many others.

The precise function of 33 Engineer Regiment (EOD) is the provision of EOD cover in peace and war for the United Kingdom and BAOR. They are not the only specialists in the EOD field for even in the Army the Royal Army Ordnance Corps (RAOC) also has an EOD responsibility and the Royal Navy and Royal Air Force also have their specialists. In order to prevent too many overlaps the roles of the two other Services are easily defined. The Royal Navy looks after purely naval ordnance such as sea mines, devices found below the high water mark and on its own property. The Royal Air Force EOD specialists look after EOD on their own property and usually close to crashed aircraft. The Royal Engineers deal with aircraft-delivered weapons while the RAOC deals with land service ammunition and leads in the disposal of terrorist-type devices, although as will be related this is not an exclusive RAOC preserve.

33 Engineer Regiment (EOD) is headquartered at Chattenden in Kent. Its full-time complement consists of two EOD squadrons, 49 EOD Squadron and 58 EOD Squadron, and the Regimental Headquarters incorporates its own Training Wing. Of the two squadrons, 49 has been established the longest – 58 was formed relatively recently. 49 EOD Squadron is rather unusual in that it has two troops manned by civilians working under Royal Engineers officers and NCOs. 1 Troop is based at Felixstowe and is still manned to an extent by Ukranians who have for years been moving around the United Kingdom clearing minefields

Below:
Saracens are used by 33 Engineer Regiment (EOD) for a variety of purposes including the safe remote control of excavators and other equipment.

Above:
The cable reel at the rear of this HYMAC excavator indicates that it is capable of being operated remotely under certain hazardous conditions.

and artillery ranges of the explosive debris of two wars. Their other civilian colleagues are 2 Troop based at Farnborough and with a similar role. 3 and 4 Troops are all-military.

58 EOD Squadron has a normal complement of three troops plus a support troop. In war they would operate in sections of eight to ten men. Working from Land Rovers with 4-tonne trucks and trailers in support they would consist of a bomb disposal officer (BDO) plus at least one senior NCO. The rest of the section would be there to provide support in the form of scouting for scattered bomblets, excavation for access to a buried device and the operation of specialist equipment. However some devices will require the attention of only one or two men and the section can if necessary be split into two or three to deal with a number of small tasks concurrently. Others might well require the support of other sections and the heavy plant and equipment provided by the Support Troop.

33 Engineer Regiment (EOD) has the following four TA Squadrons: 579 EOD Squadron (V), Dartford; 583 EOD Squadron (V), Brighton; 590 EOD Squadron (V), Rochester; 591 EOD Squadron (V), Dartford.

Mention must also be made of 491 STRE (EOD). This is formed from within the regular element of 33 Engineer Regiment (EOD) and deploys in support of the UK Mobile Force (Land).

The Royal Engineers also have their part to play in what is known as improvised explosive device disposal (IEDD), otherwise known as the terrorist device. In recent years this has become an increasing hazard to the extent that there is now an inter-service organisation known as the Joint Services EOD Operations Centre (JSEODOC). If a call for the disposal of an IED arrives at the office of the JSEODOC the team nearest to the source of the problem is sent to deal with it. Thus a team from any of the three services may find itself dealing with an IED at short notice. 33 Engineer Regiment (EOD) has a team on immediate standby at all times with another back-up team on three hours notice. This is in addition to their normal duties which includes a duty bomb disposal officer and section, a specialist diving and search team on standby, all kept within reach by radio paging.

Needless to say the regiment is kept busy and it has a presence in the Falklands clearing away the debris of the 1982 campaign. However World War 2 and even World War 1 still impose their workloads for bombs from both wars keep turning up in odd places and usually at inconvenient times. Every new 'find' sets its own standards of

difficulty and awkwardness to the extent that improvisation and ingenuity have frequently to be called upon by the personnel involved. Some devices can be rendered safe on the spot. Others have to be moved away to a place where they can be detonated. With some finds a few minutes work are all that are required, while others demand days of hard and dirty work. There is no set pattern to EOD work although over the years various tools have been devised and produced to assist in the task. Typical of these 'tools' is the Trepanner, a rotary cutter driven remotely by hydraulics or compressed air that cuts through the metal casing of a device in order to introduce steam hoses to steam out the device's contents. Various forms of microphones have been devised for use as stethoscopes to listen to any internal workings of a device and powerful electro-magnets are used to arrest the movements of timing devices. Rocket-driven wrenches are used to remove fuses or bomb pistols and so on. These devices are over and above the various items of plant used by the regiment, some of it very specialised such as the armoured dozer and the remotely-operated excavators. Then there are the various deep hole digging devices that are sometimes called upon to reach devices buried deep in the earth, and so on. For terrorist devices the regiment operates a number of 'Wheelbarrow' remotely-controlled vehicles that can be used to carry out all manner of tasks from blasting open suspect packages with a shotgun to carrying television cameras for the close examination of devices. Add to this motley array such things as water pumps to keep digging teams dry and the list of equipment used by 33 Engineer Regiment (EOD) becomes very long.

In time of war the EOD teams would assume their tasks in both the United Kingdom and BAOR. Officers are posted to every District Headquarters in the United Kingdom to act as EOD watchkeepers and to advise as required.

Above:
A special EOD dozer with an armoured cab; to the left is a small towed backhoe unit.

Above right:
A Bedford 4 x 2 TK truck with a mounted drilling rig, one of several used by 33 Engineer Regiment (EOD). *RSME*

Bottom right:
A Wheelbarrow Mk 7 used by the Regiment for IEDD duties. *MoD*

01
CP
68

Left:
A rocket wrench kit for the removal of fuses or bomb pistols with one of the wrenches fitted to a training bomb.

Below:
A microphone stethoscope fitted to a training bomb.

Above:
An EOD team at work digging a shaft to reach a deeply-buried device. *RSME*

Top:
One of the Royal Engineers' oddest vehicles, a Bombardier Muskeg tracked tractor used in locations such as the Falklands where low ground pressure is required. *Rolba*

Above:
As with so many UK-based regiments 33 Engineer Regiment (EOD) makes use of 4-tonne 4 x 2 Bedford TKs.

Left:
Car, Utility, namely a Chevette used by 33 Engineer Regiment (EOD).

7 The Postal and Courier Service

The origins of the military postal services date back to the 17th century. In 1799 the Postmaster General was requested by the Commander-in-Chief of the Army to provide an 'intelligent clerk' to accompany the Army to Europe 'to facilitate delivery and to collect letters', and a man was duly appointed. However it was not until the Peninsular War of 1808-13 that a properly organised Army Postal Service was established with an office in Lisbon. In 1882 Queen Victoria authorised the raising of an Army Postal Corps. A detachment of this corps served with the Expeditionary Force in the Egyptian campaign. During 1884 a second Army Postal Corps was formed with the title of the Royal Engineers Telegraph Reserve which was a forerunner of the Royal Corps of Signals. In 1908 the Army Postal Corps along with the Telegraph Reserve were amalgamated into the Royal Engineers (Postal Section) which went on to provide postal support to the Army during both world wars. However since World War 2 the Postal and Courier Service (PCS) has become responsible for providing postal and courier support to the Army, the Royal Navy and the Royal Air Force.

The present organistion is known as the Defence Postal and Courier Services and operates under its director, who is a brigadier, from a small directorate within the Ministry of Defence located at Whetstone in North London. Nearby is the

Below:
General view of the interior of the Mill Hill sorting office. *MoD*

Postal and Courier Depot (PCD) at Mill Hill through which all mail for the British Forces Post Offices (BFPO) passes for processing before despatch to their far-flung destinations. Within the Depot are the Postal Wing, including the main sorting office and all its various supporting ancilliary branches, an Admin Wing, a Training Wing (including the Central Volunteer Headquarters (CVHQ) to administer and train the TA members of the Service) and a Courier Wing. The Courier Wing provides a courier service for the Ministry of Defence and various Government agencies and is manned by about 40 defence couriers, all of them recruited from retired officers and senior NCOs. They operate throughout the United Kingdom and all parts of the world and often dovetail with the Queen's Messenger Service to the extent that duties are shared between the two Services. These couriers do not just carry the expected black bags containing documents. At times they are responsible for all manner of equipments and odd loads, sometimes weighing tons.

Outside the depot the service may be divided into the units within North-West Europe and units in the rest of the world. In North-West Europe, and BAOR in particular, there is 1 Postal and Courier Regiment RE, which is based in Hannover and is responsible for providing PCS support to the 1(BR) Corps units. There is also a troop in Berlin. 3 Postal and Courier Depot RE based in Düsseldorf is the main sorting depot for BAOR and additionally provides a service for all units in the Rhine area, including the RAF's 'clutch' airfields. Although not part of BAOR the small PCS Headquarters staff at Rheindahlen also controls the operations of 40 (NATO) Postal and Courier Unit RE at Casteau, south of Brussels, as part of Supreme Headquarters Allied Powers Europe (SHAPE). This unit provides postal and courier facilities for the British elements of all international headquarters throughout Europe from Norway to Italy and Portugal.

In the United Kingdom the main regiment is 2 Postal and Courier Regiment RE with its base at South Cerney. This regiment is the most ubiquitous of all the units within the Royal Engineers for it has two of its squadrons in the United Kingdom (one with detachments in Kenya and Goose Bay in Canada), one in Northern Ireland, one in Cyprus and one in Hong Kong. To these squadrons can be added detachments in the Falkland Islands (where they also provide a postal service for the local Islanders), Ascension Island,

Below:
Letter sorting at Mill Hill. *MoD*

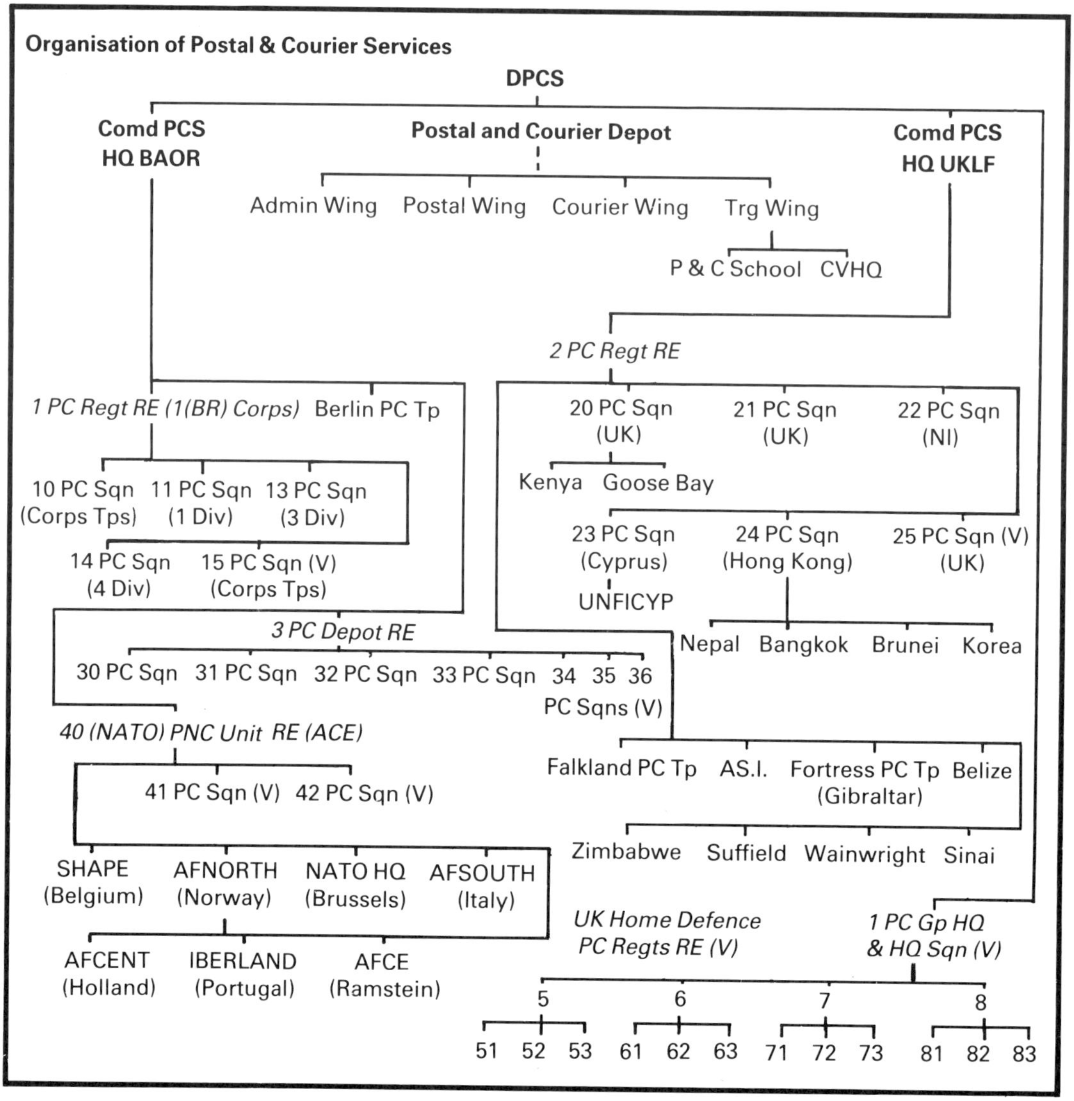

Gibraltar, Belize, Zimbabwe, Suffield (Canada), Wainright and the Sinai. The Hong Kong squadron (24 Postal and Courier Squadron) has detachments in Nepal, Bangkok, Brunei and Korea. Thus although the Regiment is based in the United Kingdom it actually operates in virtually every part of the globe and if any sudden requirement arises, such as the sending of a British contingent to the Lebanon in 1982-83, the Regiment will provide a PCS detachment to accompany it.

To all these regiments and squadrons can be added no less than four TA volunteer regiments to reinforce the existing United Kingdom PCS network in an emergency. There are also an additional four TA squadrons allocated to BAOR and two to SHAPE. The United Kingdom TA regiments have an integral presence of Royal Navy reservists and the PCD has at all times a Royal Navy manpower allocation to assist in the sorting and distribution of mail to Royal Navy Ships, a responsibility that the Royal Engineers PCS took over from the Post Office in 1962. The depot also has a strength of about 200 Women's Royal Army Corps (WRAC) personnel who are an integral part of the PCS. There are also over 200 civilians working within the Service. One small item to note is that the TA regiments and squadrons recruit almost entirely from within the civilian Post Office.

Letters and parcels posted in the United Kingdom and bearing the address BFPO SHIPS or a BFPO number are moved to London by the Post Office and arrive at one of the main line stations or a main London post office. From these points members of the PCS collect the mail daily and take it to the Mill Hill Depot for processing and sorting. From the depot the processed mail is despatched daily to Heathrow and other civilian and military airports to connect with civilian and Royal Air Force flights to their various overseas destinations. At the destination it is met by a member of the PCS and after being sorted in the Force Post Office (FPO) the mail is collected by unit representatives for the final delivery to the individual.

The sorting operation at Mill Hill is a constant process with the staff working all round the clock on eight-hour shifts, seven days a week. To add to their daily workload the PCS operates, in conjunction with various commercial mail order firms, a delivery service of goods to Service families. The mail order forms deliver their packages direct to Mill Hill in containers from where they are distributed in the normal way. The PCS makes a commercially viable charge for this service.

The various BFPOs all over the world provide a normal post office service to all Service personnel and their families and thus supply not only the usual items such as postage stamps but postal orders, saving certificates, GIRO bank facilities, Premium Savings Bonds, and so on. All these many items are delivered through the Mill Hill Depot via the accountant who is responsible for the issue of stock and the consolidation of accounts of the FPOs worldwide. The depot also has to keep its eye open for suspicious items particularly in these days when the letter or parcel bomb is an established terrorist weapon. Thus the depot is equipped with monitoring devices that are used to examine suspect mail.

On a more peaceful note the depot also runs a philatelic service for the three Services. In recent years this has become a useful method of earning funds for all manner of Service activities and the depot has special franking machines for the operation. An offshoot of this activity is a PCS museum.

Below:
Packages arriving at Mill Hill for sorting. *MoD*

Sappers joining the PCS do so as either adult recruits or junior entrants. The juniors join via the Junior Leaders Regiment RE at Dover where they complete their combat engineer training. Adult recruits are also trained as combat engineers before they move to the Training Wing at Mill Hill. Some officer recruits come from the Post Office as well as direct from civilian life and after Sandhurst they too are trained at Mill Hill.

Below left:
One of the devices used to examine suspect packages or letters at the Mill Hill Depot. *MoD*

Below right:
One of the special franking machines used for the various Services' first day or special postal covers. *MoD*

Bottom:
The final stage – mail arriving at a British Army post serving with the United Nations in Cyprus. *MoD*

8 Survey

The Royal Engineers have long had a role to play in mapping for military purposes. From medieval times it was customary to prepare plans of fortified positions and involvement in topographic surveys and map production began in 1747 when, under the Board of Ordnance, Engineer Extraordinary Watson and Practitioner Engineer Roy were sent to Scotland with a small team of draughtsmen to survey the country as part of the campaign to pacify the Highlands following the Young Pretender's Rebellion of 1745. These two duly completed their task and unwittingly became precursors of the present day Ordnance Survey when in 1791 the value of a national map of the United Kingdom was recognised. The Ordnance Survey is now a civilian organisation but it was at one time very much a military and therefore a Corps of Royal Engineers involvement.

The military have always required maps that differ from the usual civilian product and have thus taken to producing their own. Rationalisations have led to the point where the Royal Engineers are responsible for the production of maps not only for the Army but for the Royal Air Force and, in part, for the Royal Navy but for them the Admiralty continue to be responsible for the production of hydrographic charts and related products. To this inter-service end the Royal Engineers have a major-general who is Director of Military Survey with his headquarters at Feltham in Middlesex, not far from London Airport. His remit is simple for 'The mission of Military Survey is to provide the Armed Services with geographic support and related materials required for military operations on, over and in the vicinity of the land surface'. Note the use of the term 'geographic support', which these days may mean much more than just map production.

The director has at Feltham a small headquarters staff overseeing the organisation shown in the accompanying organisational 'tree'. Reporting to him are two units, the Systems and Techniques Unit and the Production Planning and Control Unit. These two units are very much forward-looking with the Systems and Techniques Unit looking forward to the surveying and map-making techniques that will be used in the years ahead. Recent years have seen a tremendous

Organisation of Military Survey

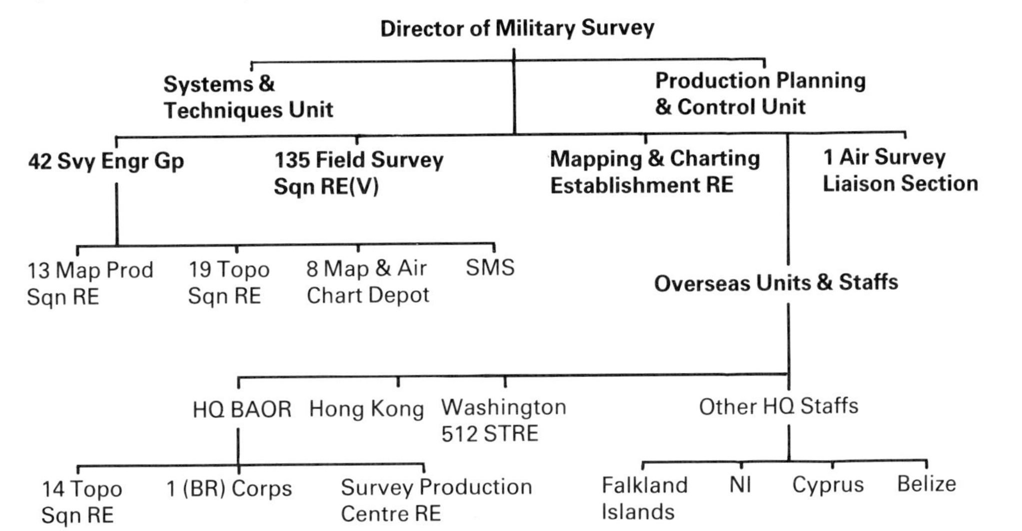

Above:
A survey team in action in the field. *Soldier Magazine*

upheaval in the technology of all stages of map production with computers and their related digital technology revolutionising methods but the revolution is not yet over. Navigation satellites have already been drawn into the system for positioning purposes and it is anticipated that developments in imaging satellites such as LANDSAT may be bought into the mapping process in the future. Exactly how this will be done is the province of the Systems and Techniques Unit. The Production Planning and Control Unit is more of a supervisory unit that plans the work of the Directorate to make best use of its resources to meet military requirements. It does not work in isolation for in order to ensure standardisation and to avoid duplication of production of maps and charts NATO has assigned areas of responsibility to the various member states. The United Kingdom's area of interest extends outside the NATO land area and the United Kingdom is thus concerned not only with producing maps of the United Kingdom and the operational areas within Europe but also engages in collaborative surveys with some of the Commonwealth nations and other countries.

The director commands a far-flung organisation but before dealing with the more outlying areas the United Kingdom units are best considered first. The two largest units are 42 Survey Engineer Group and the Mapping and Charting Establishment RE, or MCE.

42 Engineer Group has its headquarters at Hermitage, near Newbury, although two of its main units are 13 Map Production Squadron RE and 19 Topographic Squadron RE at Barton Stacey, just north of Winchester. The two other units in the Group are 8 Map and Chart Depot at Guildford in Surrey and the School of Military Survey, again at Hermitage. These units are the 'military side' of Military Survey in the United Kingdom. The MCE is very much a civilian-manned organisation, some 600 strong with half of this strength made up by cartographic draughtsmen. The MCE is located at Feltham and acts as a form of map production 'factory' that can carry out all stages of map production from basic inputs to the finished product. The actual stages of map production are fascinating to a high degree but to go through them all in detail would fill this book. Using information from a variety of sources the MCE can rapidly produce maps of many parts of the world in various scales. The inputs for this operation are varied and range from the use of field surveys to aircraft photographs. The surveys themselves are fascinating for many of them were and still are produced by the time-honoured method of a surveyor journeying across uncharted regions and using survey points and instruments to plot areas and report back, thereby continuing one of the established images of the Royal Engineers, ie an individual in a pith helmet peering through a theodolite. These survey reports are filed by the MCE in a section known as the Map

Below:
Map printing using one of the large Heidelberg presses held by 135 Field Survey Squadron (V) at Ewell. *Soldier Magazine*

Research and Library Group that contains not only the surveys but maps of all kinds from all parts of the world, ranging from city guides to street maps. Other sections in the MCE include the Computer and Geodesy Group wherein the mysteries of the finer points of survey are practised and the Production Group where the maps are actually made, including such oddities as the 'moving map' transparencies for use by the Royal Air Force's Harrier and Tornado squadrons. The MCE also has the Technical Services Group to support the other MCE operations.

The MCE makes much use of computers and all manner of technical wizardry to produce their maps and other products and much use is now made of photographic reconnaissance to cover the many areas of the world still required for survey or up-dating purposes. The photographs are mainly taken by the Canberra PR9s of 1 Photographic Reconnaissance Unit (1 PRU) based at RAF Wyton. To liaise with the Royal Air Force and to assist in mission planning and general co-ordination of efforts Military Survey maintains 1 Air Survey Liaison Section at RAF Wyton.

Also part of the United Kingdom organisation is 135 Field Survey Squadron RE(V), a TA squadron based at Ewell in Surrey. Although a TA squadron this unit carries out all manner of survey duties and was, for instance, responsible for the production of a map of Alderney, one of the Channel Islands.

The directorate maintains various staffs in several overseas locations, with one of the largest representations being in BAOR. Apart from a staff at HQ 1(BR) Corps there is also a Survey Production Centre at Mönchengladbach. Ratingen, just north of Düsseldorf, is the base of 14 Topographic Squadron, a mobile unit which was the first to use TACIPRINT, a method of rapidly printing formation and other symbols onto existing maps for operational purposes down to company level. TACIPRINT is now used by other units within the Directorate.

A unique unit of Military Survey is 512 STRE based in Washington DC. This team works closely with the American Defense Mapping Agency in improving world knowledge of the physical characteristics of the earth by Doppler observations of space satellites. They have worked in over 60 countries in the last 10 years providing survey data for the local mapping organisations.

Other overseas staffs are maintained in Hong Kong, the Falklands, Cyprus and Belize. Closer to home a staff is maintained in Northern Ireland. These staffs provide map and chart stores and also continue their survey duties as and when required.

At present the Director of Military Survey commands an organisation where the 'military' personnel are outnumbered by their civilian counterparts but in the main it is the survey officers and technicians who actually venture forth into the far-flung and remote regions and carry out the in-the-field surveys. Operationally it is the Royal Engineers who produce the required maps and charts, store them and distribute them as required. They also have a considerable forward-looking responsibility to all three Services for often maps have to be produced in quantity in a great hurry. The experiences of Operation 'Corporate' are only a few years old and at that time the MCE and 42 Survey Engineer Regiment (as 42 Survey Engineer Group was then) worked around the clock producing all manner of maps and charts, not only of the Falklands but also of all the likely en route localities, both in Africa and South America. A survey presence is still maintained in the Falklands, complete with a TACIPRINT facility.

Above:
A Sapper surveyor using one of the new Kern K1 theodolites coupled to a DM502 electro-optic distance meter. *Survey & General*

9 The Military Works Force

The Military Works Force (MWF) is one of the branches of the Royal Engineers least known to the general public but it has considerable achievements to its name despite its relatively short history. In its present form it can trace its origins back to 1959 when many of the various works responsibilities carried out by the Corps were passed to the Ministry of Works, now the Property Services Agency (PSA), a part of the Department of the Environment. Despite this handover it was decided that the Royal Engineers would retain some works responsibilities for areas where the PSA organistion was unable to operate for various reasons, the main ones being where no organisation had ever existed or had been destroyed by conflict, as happened in the Falkland Islands in 1982. In 1965 special teams were formed by the Corps as the result of operational experience in Borneo and Aden and from these teams evolved the current Military Works structure.

The Military Works Force in its present form was established in 1978 and until September 1982 it formed part of the Engineer Support Group. On that date as the result of an internal reorganisation it was moved to become a unit of the newly re-formed 12 Engineer Brigade (ADR) (see Section 6). The MWF is now a unit within this brigade, although it is sponsored by the Ministry of Defence and tends to operate independently.

The primary role of the MWF is to provide an engineering consultancy to all three of Her Majesty's Forces in the field. In time of peace this is a PSA function but the role would be assumed by the MWF in wartime, during military operations or whenever it is deemed necessary for reasons of safety, security or speed. In addition to this the MWF also provides technical support to other Royal Engineer units when they are on either military or disaster relief operations. MWF also carries out a design and supervisory role for Royal Engineer projects that are undertaken annually by the Corps in various parts of the world, including the United Kingdom. These include large scale construction projects that are carried out partially

Below:
Part of the fuel installation designed and planned by the MWF on Ascension Island during Operation 'Corporate'. *MWF*

to offset the costs involved in the use of training facilities in countries such as Kenya and Canada, but at least one major project and several more minor ones are also carried out annually for the PSA within the United Kingdom. Apart from the consultancy involvement the MWF frequently carries out the design and supervisory role.

Above:
The fuel pipeline built under MWF supervision on Ascension during the early stages of Operation 'Corporate'. *MWF*

Below:
One of the radar installations in the Falklands designed and planned by the MWF and constructed by the Royal Engineer work force. *MWF*

The MWF also has an important function in supervising the maintenance of various essential services in some sensitive installations in the event of an emergency. These services are extensive and some are rather too sensitive for open discussion, but some that can be mentioned include the running of power stations, the construction of emergency fuel installations and the maintenance and repair of railways. The MWF also supplies specialist technical advice to all the armed forces and can also advise overseas services. If all this were not enough the MWF is sometimes able to provide services to various civilian development agencies and to some charitable organisations.

Above:
The *Pursuivant*, built under MWF supervision at Lowestoft to provide floating 'Coastel' accommodation for the troops in the Falklands. *MWF*

Below:
Mount Pleasant airfield in the Falklands, built by a civilian consortium following site reconnaissance by the MWF. *MWF*

To carry out these various roles the MWF is a

Above:
Building a small arms range at the RMA Sandhurst during 1981, a MWF project for the PSA. *MWF*

Ministry of Defence sponsored unit and is tasked by the Director of Engineer Services at the Ministry of Defence in Whitehall through 12 Engineer Brigade. Headquarters Military Works Force is at Barton Stacey just to the north of Winchester in Hampshire. The Commander Military Works Force has under his command a headquarters staff and two units known as 62 CRE (Works) and 64 CRE (Works).

The headquarters has the normal staff functions common to any military unit plus some unusual ones. It deals with the normal clerical, quartermaster and transport aspects but in addition to this important function there are within the HQ five specialist departments or cells. The first of these is the Construction Materials Laboratory with a wide range of very specialised and sophisticated equipment for obtaining and testing all manner of materials, ranging from soils to the raw materials used for construction. There is also the Procurement Cell responsible for the actual procurement, movement and accounting of all manner of construction stores, often at very short notice, so this cell maintains a close liaison with the construction supply industry. The Estimating Cell has responsibility for the checking and collating of estimates of costs, labour and materials. There is also the Report Production Cell which produces and publishes the engineer reconnaissance and planning reports generated within the MWF, and last but not least, there is the Computer Cell with facilities for graphics and computer-aided design for the support of all the teams within the MWF. All of these cells are backed by a Technical Information Centre and Library which is also available to the rest of the Corps.

The term CRE is an old one within the Corps and denotes Commander Royal Engineers and it has been traditionally used in the Works Service organisation. Each of the two CRE (Works), 62 and 64, has its own headquarters, a drawing office and a survey team. Within each CRE (Works) are three Specialist Teams Royal Engineers, or STREs. Each STRE can form a number of design teams from the four officers and eight technicians who are either Warrant Officers or Senior NCOs. Each STRE can deal with two major projects at any one time. Four of the STREs involved with the regular MWF are Works teams, one is a specialised Well Drilling STRE and the other is a Bulk Petroleum STRE. Their organisation can be seen on the accompanying chart. In wartime the MWF moves to Germany and takes over from the PSA. It is reinforced by 522 STRE (Works) which is normally based in Germany at Willich. In addition there are a number of individuals from Military Engineering Services (Works) who are Royal Engineers attached to the PSA. At all times the MWF can call upon the expertise of the 16-strong Engineer Specialist Pool manned by former STRE commanders; as one would expect, their range of expertise and knowledge is considerable.

To back up these regular MWF units are a number of Territorial Army STREs. Their designations and specialist skills are given in the accompanying table, but it should be stressed that most of the personnel involved are civilians with skills and knowledge that will be used by the MWF in an emergency. An example of this can be seen with the three STREs (Bulk Petroleum) who are made up of personnel from ESSO, BP and Shell respectively. In time of peace these STREs are administered by the Central Volunteer Headquarters Royal Engineers (CVHQ) but in war or in a period of tension they are under MWF command and most of them would travel to Germany to maintain essential services in British installations according to their specialisations. Some PSA personnel already working in Germany also have a TA commitment and simply don their military uniforms and continue their duties with MWF.

List of TA STREs

501 STRE (Bulk Petroleum) (V)
502 STRE (Bulk Petroleum) (V)
503 STRE (Bulk Petroleum) (V)
504 STRE (Works) (V)
506 STRE (Works) (V)
507 STRE (Railway Construction) (V)
508 STRE (Works) (V)
520 STRE (Well Drilling) (V)
525 STRE (Works) (V)
526 STRE (Works) (V)

Thus to sum up the MWF can provide a wide range of engineer services including general advice, feasibility studies, project reconnaissance, a complete design service, project planning, technical supervision and project management using military or civilian labour. Specialist services include engineering surveys and inspections, site investigations, materials laboratory services, well drilling, the design and construction of military fuel installations and the provision of technical information.

This wide range of services has been put to use in many ways since 1959, usually escaping public attention. Nevertheless the MWF has acted as a quiet ambassador all over the world. No project they have been involved with has outstripped the efforts involved in the Falkland Island campaign of 1982 and the aftermath. The MWF was involved with the finer points of the planning of Operation 'Corporate' from the beginning and only four days after the invasion of the islands by Argentina a reconnaissance party from the MWF was on Ascension Island to assess what would be required to transform a quiet backwater into a major military forward base. One of the subsequent major construction projects involved a bulk fuel installation that included 3½ miles (5.63km) of pipeline, constructed in 3½ days by a Royal Engineers Field Squadron (Construction) Troop supervised by members of 516 STRE (Bulk Petroleum). Other installations on Ascension included a reverse osmosis water purification plant, air-conditioned airfield accommodation and the rapid expansion of a tented camp along with all the necessary services.

MWF personnel sailed with the Task Force (on the QE2) and during the actual campaigning they established a fuel system for Port San Carlos to supply aviation fuel for Harrier and helicopter operations (at one stage delivering at the rate of 40,000gal a day). Even as Port Stanley was being mopped up MWF personnel were busy initiating immediate repairs to the water and electrical services and surveying the state of the runway at the airfield. It was at this stage that the early planning paid off as one decision had been to establish a floating 'builder's yard' containing all manner of construction materials and equipment such as generators. It was originally planned that these materials would be used for garrison accommodation in the form of two 800-man and seven 200-man hutted camps complete with power, water and sewage services. Every item had to be planned in minute detail for any omission would be discovered on the end of an 8,000-mile (12,870km) long logistic chain. As it turned out the planned hutted camps were not required and instead 100- and 400-man camps were constructed and sized to house specific units. As can be imagined this involved a great deal of MWF planning 'on the spot' and the floating 'builder's yard' proved to be invaluable. In all seven hutted

Above:
One of the first items of equipment taken to the Falklands was a rock-crushing plant similar to this one used for training at Chattenden. It provided much raw material for MWF projects in the Islands.

camps were built at a cost in materials alone of over £5million.

Apart from this purely local task much of the MWF planning and design for the Falklands installations was carried out in the United Kingdom, often in close collaboration with industrial concerns. In some cases, such as large radar installations and the associated sites, the MWF was responsible for the design, planning and supervision of the project. In others, such as the new airport on Mount Pleasant, the MWF was responsible for the reconnaissance of the site and the feasibility study. The detail design and control was taken on by the PSA with the construction being carried out by a civilian consortium.

Not content with the design and planning of accommodation and airfields the MWF has also been involved in the subsequent provision of three floating Coastels (coastal hotels) two of which were leased from a Swedish concern. The third, named the *Pursuivant*, was built by a Lowestoft shipyard under MWF supervision. All three of these Coastels have containerised accommodation built onto a 90 x 30m North Sea barge and contain cookhouses, messes and even some sports facilities. They have their own power supply systems, water desalination and sewage installations and with the *Pursuivant* the MWF supervised it all – with the Swedish Coastels MWF involvement was limited to delivery checks.

To these Falkland Island projects can be added numerous minor projects such as the provision of laundries, workshops and other such essential facilities. A new semi-automated power station has been provided to replace the emergency station in use since the conflict (that too was designed by the MWF), and various services have been planned for the radar stations including proper access roads and water supplies. A new ship-to-shore installation has been constructed for the Port Stanley airfield to replace that built by the Argentinians (who also provided a great deal of the aluminium matting used in the initial repair of the runway).

At one time or another nearly all MWF personnel have been involved on site in the Falklands, usually on five-month tours. The number of Falkland Island projects carried out by the MWF now runs to several hundred and while the bulk of the work is now complete or nearing completion, there are still some continuing tasks to plan, design and supervise, not only in the South Atlantic but elsewhere.

One of these is well drilling. 521 STRE (Well Drilling) is the only regular unit of its type in the British Army and is usually kept active by a continuing programme of well drilling to provide emergency water supplies for Royal Air Force

Above:
One of 521 STRE (Well Drilling)'s Magirus-Deutz rigs next to a towed Dando rig.

airfields. For this job the STRE has a number of towed Dando drilling rigs and a number of larger self-propelled drilling rigs carried on Magirus-Deutz 6 x 6 truck chassis. A typical example is the well at RAF Coltishall which is 79.5m deep and yields 2,000cu m of potable water every day – ten times the daily requirement of the station.

Firing range design is a fairly common MWF task and recently MWF personnel have been involved in the design of a small arms range on Gibraltar and a close quarter battle range in the UK, the latter being for the PSA. Another recent range project for the PSA was the design of a 600m range at the Royal Military Academy at Sandhurst, but perhaps the largest of these training projects is the design and planning of a complete School of Infantry for the Kenyan Army at Isiolo in Northern Kenya. Known as Exercise 'Larch Pole' this is a seven-year project that commenced during 1982. As always the MWF is responsible for the design and planning of everything from ranges to messes and from water towers to roads, but the actual construction is carried out by various Royal Engineer field squadrons. The programme is part of an arrangement made to compensate for the use of various training facilities provided for the British Army in Kenya. Similar arrangemens exist elsewhere, and in Canada Exercise 'Water-Leap' is set up on a similar basis to the Kenyan operation – the MWF was responsible for the design and planning of the Petawa Bridge in Ontario. This is a 64m class 60 bridge capable of withstanding the rigours of the Canadian climate whose construction opened up the access to a nearby Canadian Armed Forces base. The latest 'Canadian' project involves the construction of an ammunition depot at Cold Lake, Alberta. Norway is another recipient of MWF design and planning expertise – Rjoandani Bridge was designed by the British and when built during 1981 which involved the use of Sea King helicopters to place some of the steel beams.

The various TA STREs also find themselves involved in similar projects. The Bulk Petroleum STREs have an involvement in the fuel line replacement programme in the Gibraltar dockyard area and other STREs have been involved in water supplies and other such ventures on the Rock. Hong Kong has been the subject of much TA STRE attention ranging from firefighting installations to drainage on training ranges. Nearer to home various TA STREs have been involved in the planning and construction of footbridges in remote locations and even in the planning and construction of a new electricity system on one of the Scottish islands. For the latter some support was provided by the MWF.

10 The Depot and Training Establishments

The Depot Regiment

Within the British Army every regiment and corps has its own depot which in a way acts as the formation's 'home', wherever the formation might be serving. Since the Royal Engineers are even more widely distributed than most its requirements for a home base is rather more pressing than usual and accordingly has its home base at Chatham in Kent, the location of the Depot Regiment RE.

The Depot Regiment RE has two squadrons, the Depot Squadron and 36 RSME Squadron. The latter is easily explained for it is an administrative squadron formed to administer and hold soldiers attending the Royal School of Military Engineering (RSME) at Brompton Barracks. The Depot Squadron is rather more involved for it carries out many of the usual depot functions required of its establishment – these will be mentioned below – but there are other functions carried out by the Depot Regiment as a whole.

The Commanding Officer of the Depot Regiment RE commands not only the two squadrons but also the various messes and the Royal Engineers Staff Band, now the only one used by the Corps; the Aldershot Band was a victim of defence spending cuts. Of the messes the most important is the Royal Engineers Headquarters Officers Mess, the Headquarters Mess for the entire Corps. It is resplendent with an array of trophies, relics and effects of all kinds that mark the history and duties of the Corps over the years. Like the Depot Regiment itself, this mess is located within Brompton Barracks.

The functions of the Depot Regiment are many. Top of the list has to be the rather mundane day-to-day function of simply holding and administering the officers and other ranks based in Brompton Barracks and conducting their routine military training. There are also a number of civilian staff members who have to be administered and the security of the barracks area is a sizeable task as well, to say nothing of the maintenance of the many areas and structures within the general confines of the barracks. A sizeable stock of technical stores and equipment is held by the regiment and a recent audit revealed that this equipment had a surprising value of over £22million. Some of this equipment is used for the carrying out of the depot functions which apart from the usual holding of the various odd bodies 'between postings' and the like involves the holding of the kit of personnel on extended

Below:
Recruit undergoing demolition training at 3 Training Regiment (RE) and using 'hayrick' charges.

Above:
Sapper corporal carrying out basic tool training for assault pioneers at 3 Training Regiment.

postings in distant locations and the holding of the kit and effects of casualties. The depot is also responsible for the processing of personnel leaving the Corps and the Army and for this function alone about 700 pass through the Depot Regiment every year. The Depot Regiment is also used to 'hold' various personnel who are carrying out 'one off' staff and other duties.

The Depot Regiment RE carries out various training courses for the Corps as a whole. Every year the depot runs the Royal Engineers Tactics Refresher Course for officers who are taking the examination for entry to the Staff College, Camberley. There is also the Warrant Officers Commissioning Course held every year for some Warrant Officers who are commissioned from the ranks to carry out duties of a technical, administrative and logistic nature. Courses are also carried out for RSMs and senior NCOs, usually dealing with man management. The depot also runs briefings for personnel who have applied to the Regimental Commissioning Board (RCB) for commissioning.

Before leaving the Depot Regiment RE it would be as well to mention that it comes under the command of the Headquarters RSME as does all the Chatham Garrison, which includes units in Maidstone. HQ RSME is also responsible for the RSME itself, 12 RSME Regiment RE (of which more below), and 69 Gurkha Independent Field Squadron, also based at Chatham. This Queens Gurkha Engineer Squadron is not part of the Brigade of Gurkhas and in an emergency would support 2 Infantry Brigade. It is also worthy of mention that in an emergency the depot personnel would probably be dispersed amongst various field squadrons, including those involved in home defence, leaving only a small nucleus to keep the depot functions running.

The Training Establishments

The Royal Engineers have several training establishments, some of them very specialised (such as the School of Military Survey at Hermitage, near Newbury), but this section will deal only with those that 'process' the Corps' soldier intake. For officers that means part of the RSME at Chattenden in Kent and for the soldiers the training establishments of 11 Engineer Group. These are 1 Training Regiment RE, 3 Training Regiment RE, the Junior Leaders' Regiment at Dover and the Army Apprentices College at

Chepstow. Before dealing with the actual training it would be as well to outline how a Sapper is trained from the time he joins the Corps.

Initial recruit training is carried out by 1 Training Regiment where the usual 'square bashing' and basic military skills are taught. The recruit then passes to 3 Training Regiment where the combat engineering skills are imparted. At the end of this the trained Sapper passes to his first posting as a Combat Engineer Class 3 for further training in a field unit. After about a year the training carried out within his unit will usually raise him to the level of Combat Engineer Class 2. To become a Combat Engineer Class 1, ie a combat engineer capable of technical supervision of engineer tasks, the Sapper will usually be a NCO. By that stage the Sapper will also have been provided with some form of skill training for a fully trained combat engineer is also a craftsman in the form of a metal worker, electrician, welder, carpenter, and so on. This Class 1 level training is carried out by 3 Training Regiment or in Germany. It is perfectly possible for a Class 1 Sapper to be just a Sapper but some are raised to NCO rank by the time they have reached that stage. In any case a Class 1 Sapper has to have some supervisory experience.

The above is only a very simplified outline of the training structure that a Sapper might take. It has not even mentioned the Junior Leaders or the Army Apprentices who are provided with a great deal of training even before they join adult service. The main difference between the two is that the Junior Leaders are given some of their combat engineer training while they are still with the Junior Leaders' Regiment. The Army Apprentices are given their combat engineer training by 3 Training Regiment. Other differences exist as well, but at the time of writing it seems possible that these two formations will become merged once the planned move of the Junior Leaders' Regiment from Dover to Chepstow takes place, probably in the late 1980s.

1 Training Regiment is at Gibraltar Barracks near Minley and Hawley, and just north of the home of the British Army at Aldershot. It is not just a recruit training establishment for it is a complex unit, but its main task is the introduction, administration and training of recruits for the Corps, both regulars and TA. Junior NCOs for the TA are also catered for and 1 Training Regiment also trains its own instructors (and those of 3 Training Regiment as well). It also trains the Corps' Assistant Instructors

Below:
Basic MGB training with manpower much in evidence.

Above:
A crane belonging to 56 (MT) Training Squadron of 1 Training Regiment being used for driver training.

Physical Training (AIPTs) before they get their specialised training at the Army School of Physical Training in nearby Aldershot after which they return to tone up the Corps. Pre-course training is also provided for Corps NCOs attending various courses such as the Warminster skill-at-arms course. Various forms of training assistance are provided for Army and Combined Cadet Force units in the general locality of the regiment and if all this were not enough it is responsible for the highly specialised training of basic HGV 3, 2 and 1 drivers and crane operators for the many vehicles within the Corps inventory.

The Regiment's Headquarters administers three squadrons, 27 Headquarters and Volunteer Training Squadron, 28 Training Squadron and 56 (MT) Training Squadron. RHQ also administers the Chief Instructor's Department, the usual Quartermaster functions, a Medical and Dental Centre and the Training Regiment RE Workshops, REME.

The bulk of the regular recruit training is carried out by 28 Training Squadron with five training troops under the squadron headquarters. Each troop has four sections, each under a junior NCO and usually containing 12 recruits. The basic course lasts 10 weeks involving all the usual drill, PT and basic military training that would be expected. Facilities available for the training are first class with plenty of sports facilities and other methods of converting civilian flab into military

Below:
Watermanship training using a Mk 5 assault boat.

muscle. Some of these facilities are also used by TA recruits whose basic training consists of 15-day non-stop courses carried out by 27 Headquarters and Volunteer Training Squadron.

The driving instruction carried out by 56 (MT) Training Squadron is divided between three troops, the Basic Driver Troop (BDT), the Advanced Driver Troop (ADT) and the Specialist Driver Troop (SDT). The BDT training involves all the basic driving skills plus some military ones, such as vehicle camouflage and is mainly carried out using 4-tonne trucks. ADT trains drivers to Class 2 level – ie those able to practise Class 1 Driver training – and the SDT instructs in the use of cranes and other specialised driving. Some driver training is also provided for other corps using similar equipment; much of this is provided by civilian instructors.

The general day-to-day administration of Gibraltar Barracks is shared by 3 Training Regiment who also use the premises. The latter may be regarded as the combat engineer training regiment and, apart from the usual regimental headquarters, it has three squadrons, 55 and 57 Training Squadrons and 63 Training Support Squadron which also supports the activities of 1 Training Regiment. Each of the two training squadrons has five training troops and combat engineer courses are provided not only for the Corps but also for infantry assault pioneers and Royal Armoured Corps assault troopers. The basic Class 3 courses last 11 weeks for recruits and Army Apprentices (Junior Leaders seven weeks) and Class 1 courses 10 weeks. Infantry and RAC courses last six weeks. TA training is also provided.

The basic combat engineer skills course covers what is known as EBS (Equipment – ie tools – and the Basic Skills), field defences (including the use of barbed wire and obstacles) water supplies, basic demolitions, watermanship (on Hawley Lake), seven days of mine warfare, improvised bridging (also known as NEB – non-equipment bridging), MGB and HGOB construction, concrete laying and PRA (plant, roads and airfields). The Class 3 course also includes a four-day exercise in the field involving a number of combat engineer tasks, some carried out under NBC conditions.

At the end of the course the recruit is passed out as a trained combat engineer and is posted to a field unit, but only if 3 Training Regiment considers him good enough. If he is not up to the mark for any reason he will be retained until he is considered good enough. The regiment contains its own 'quality control' system as does 1 Training Regiment, but there is also the Training Validation and Testing Department which is separate from both and responsible for overall training standards.

Not all combat engineer training is carried out at Gibraltar Barracks. Class 1 training may also be carried out in BAOR by the Combat Engineering Training Centre at Gordon Barracks, Hameln (Hamelin) where identical training to that provided by 3 Training Regiment is provided for Sappers based in Germany. Infantry and RAC training is also provided within the barracks and on the nearby training areas at Potsen and Holdensen. The centre also provides a mobile team that tours all units within BAOR and lectures them on the threat expected in time of war and the forms that the threat might take.

Back in the United Kingdom a watermanship establishment is maintained at Wyke Regis which is used not only by regular units but by the two 'Junior Entry' establishments of the Corps. As has already been mentioned these two latter establishments are likely to merge or at least use the same premises at Chepstow now used by the Army Apprentices College.

The Junior Leaders' Regiment Royal Engineers is currently at Old Park Barracks near Dover. The Corps uses both Junior Entry establishments to provide its future NCOs and in both establishments attention is given to the development of leadership potential. This is not the only aspect of junior training for, at the end of it all, the end product should be a soldier who can live in the field and remain effective with all the

Below:
Marksmanship training using a L2A3 sub-machine gun.

mental robustness, tactical awareness, self-administration, physical fitness and skill-at-arms that the task entails. The two establishments differ in some details – for instnce the accent on the Army Apprentices College entrants is trade training while the Junior Leaders concentrate on combat engineering – but perhaps this is an over-simplification. At both training is provided in all the basic military skills from square bashing to weapon training. Physical training is emphasised and sport of all kinds is encouraged to a high degree. Adventurous training is provided to build self-confidence, and hobbies of all kinds are encouraged. Educational training includes military calculations, orientation for service life, engineering sciences, communication skills, first aid and map reading as part of a syllabus leading to at least a Junior Army Education Certificate. As a general rule the standard of entry to the Army Apprentices College is higher than that of the Junior Leaders Regiment and pass-out leads direct to Class 2 standard in a trade with the full combat engineer course at 3 Training Regiment then ahead. Junior Leaders carry out part of their combat engineer training at Dover – they spend a year there – using an area at Yantlet Creek for much of the practical training.

Officers for the Corps first have to pass through the Royal Military Academy (RMA) at Sandhurst, taking the usual seven-month course. They then pass to the RSME at Chattenden for their Young Officers (YO) Course which teaches them their necessary military engineering skills. The RSME is a busy place: every year it carries out no less than 129 different types of course involving a turnover of some 6,500 men. To carry out this task the RSME has the support of the Depot Regiment at Brompton Barracks and 12 RSME Regiment consisting of 17 RSME Squadron, 18 RSME Support Squadron and the inevitable REME Workshop. There is also a field squadron, 24 Field Squadron, which is used by the RSME for general support and the provision of manpower for various purposes, including demonstrations.

While this squadron is actually based at the RSME and is part of it, in an emergency it would become part of 22 Engineer Regiment based at Tidworth and would then become part of 1 Infantry Brigade.

Many courses are run by the RSME, some of them very specialis ed. To carry out all these courses the RSME is divided into a number of wings. The Field Engineering Wing (FEW) deals with all aspects of combat engineering, including mine warfare. The Tactics Wing deals with the way the Sappers integrate and operate with other combat arms on the battlefield, and associated topics. The Signals Wing works very closely with the Royal Corps of Signals and deals with the way the Corps communicates in the field. The Plant, Roads and Airfields Wing (PRA Wing) deals with heavy plant operation and management. There is also the Electrical and Mechanical Wing and Civil Engineering Wing. Outposts are maintained at Bovington to deal with armoured engineer aspects and the Royal Engineers Diving Establishment is at HMS Vernon, Portsmouth.

All these wings have their own internal organisation to deal with the instruction of their particular specialities. Some of the courses are fairly basic for young officers while some of the more advanced courses are provided only when an officer has been in the Corps for some years. Not only officers may be involved – some of the courses are provided for NCOs and senior NCOs and others are provided for infantry assault pioneer and RAC assault trooper officers and NCOs. At times the RSME takes advantage of various civilian training establishments for special purposes or knowledge and some of the RSME staff and instructors are civilians. Some of the facilities provided at the RSME rival those at any civilian university and vary from laboratories of all kinds to technical and other libraries. Also resident at the RSME is the Doctrine and Weapons Team that deals with the future equipment and operating methods of the Corps. Co-located with the RSME is the Royal Engineers Training and Development Team (RETDT).

Left:
Pistol training on the Hameln ranges.

Others

In the course of this book almost every unit of the Corps of Royal Engineers that can be mentioned has been mentioned but even so there are a few units which, for one reason or another, do not fit into any tidy category. One of these is the Royal Engineers Sales Team, a unit that forms part of the staff of Military Defence Sales, the organisation that sells military hardware to other nations. Engineer equipment of all types features high on its sales list and consequently the RE Sales Team was formed to demonstrate this equipment to all concerned.

Mention has not been made of two lighting units with entirely different roles. In BAOR a Corps Lighting Troop is based at Hertford to provide electric power and lighting for the Corps headquarters in the field. 873 Movement Lighting Squadron RE(V) has searchlights to raise the ambient light level (artificial moonlight) for various forms of night operations, including floodlighting for bridging operations.

Apart from its involvement with 28 Amphibious Engineer Regiment, diving has hardly been mentioned but it is an important Corps function. The Royal Engineer Diving Establishment was based at Marchwood on Southampton Water but was due to move to HMS Vernon, Portsmouth, to be co-located with Royal Navy training facilities during 1985.

Military Engineering Services (MES) are part of the Royal Engineers and deal mainly with the organisation and maintenance of some 'civilian' type services at large military installations, usually overseas. MES (Works) are maintained at Gibraltar (also home of 1 Fortress STRE, the suppliers of the Rock's water supplies among other things), Berlin, Cyprus and Hong Kong.

Left:
A searchlight used by 873 Movement Lighting Squadron. *TJG*